Beam-based Correction and Optimization for Accelerators

Beam-based Correction and Optimization for Accelerators

Xiaobiao Huang

CRC Press
Taylor & Francis Group
Boca Raton London New York

CRC Press is an imprint of the
Taylor & Francis Group, an **informa** business

CRC Press
Taylor & Francis Group
6000 Broken Sound Parkway NW, Suite 300
Boca Raton, FL 33487-2742

First issued in paperback 2021

© 2020 by Taylor & Francis Group, LLC
CRC Press is an imprint of Taylor & Francis Group, an Informa business

No claim to original U.S. Government works

ISBN-13: 978-1-138-35316-9 (hbk)
ISBN-13: 978-1-03-217654-3 (pbk)
DOI: 10.1201/9780429434358

Visit the Taylor & Francis Web site at
http://www.taylorandfrancis.com

and the CRC Press Web site at
http://www.crcpress.com

Contents

Foreword

It is an honor and a pleasure to write the foreword to this overview of beam-based methods for improving the performance of particle accelerators.

The timing of this book is excellent. The new generation of particle accelerators now proposed or in construction have taken full advantage of the latest improvements in accelerator design and engineering, including magnet construction, survey and alignment, vacuum technology, RF, beam diagnostics, and lattice design, both linear and nonlinear. The result is accelerator performance pushed well beyond that which would have been reasonable to pursue only a decade ago. Advanced techniques in beam-based accelerator control will be required to commission and realize the full potential of these advanced new machines. This book gives a review of experimental techniques used to find and correct errors and maximize accelerator performance. These techniques range from well-known, basic measurements to the latest innovations in beam-based optimization and machine learning. The focus is on algorithms that efficiently get results and have been proven to make measurable improvements to the performance of existing accelerators. This book provides an essential guide for the physicists who commission and operate these accelerators.

Dr. Xiaobiao Huang is well-suited to present this material. I have had the good fortune to work with Dr. Huang for the past 13 years, since he came to the Stanford Linear Accelerator Center to work at SSRL early in his career as an accelerator physicist. He had already established a name for himself in developing beam-based accelerator control techniques in his previous work as a graduate student with S.Y. Lee at Indiana University and as a research associate at Fermilab, in particular for his work using independent component analysis (ICA) to debug accelerator linear optics. Since then, he has made many additional important contributions, including improving the orbit response matrix analysis (LOCO) so it produces more stable and reliable results for fitted accelerator linear optics; developing a number of beam-based optimization algorithms, including robust conjugate direction search (RCDS), that have succeeded in improving nonlinear optics; and, most recently, applying machine learning techniques for online accelerator optimization. Dr. Huang's work brings the full power of the latest computational algorithms to accelerator optimization. His combined expertise in experimental measurement, mathematical analysis, and computer programming have naturally led to his playing a central role in this rapidly developing field.

James Safranek

Preface

Particle accelerators have played a critical role in the history of scientific discovery. They provide a controlled means to probe the atomic and sub-atomic world by boosting charged particles to high energies and colliding them with a fixed target or another beam of charge particles. Within the last 90 years, accelerators with ever higher energy reach were built, enabling the discoveries of a series of fundamental particles and leading to the establishment of the Standard Model. Accelerator development in the energy frontier culminated in recent years, with the completion of the Large Hadron Collider (LHC) and the discovery of the Higgs boson.

On a different front, accelerators have been used to generate intense X-rays for the studies of atomic and molecular structures or processes in ordinary matter. In these accelerators, which are called light sources, high energy electron beams, under the forces of strong magnetic fields, emit highly directional, high flux photon beams. Research conducted at light source facilities is critical to solving many of the challenges that the human society faces today in, for example, the environment, energy, and medicine.

High energy particle accelerators in high energy physics or light source applications are mostly synchrotrons or linacs. Both synchrotrons and linacs are complex and delicate machines. The geometric footprint of a synchrotron or a linac can vary from meters to kilometers, while the required precision of beam control is typically sub-micron. A machine is often composed of hundreds or thousands of components, such as magnets, RF cavities, vacuum pumps, kickers, and diagnostic pick-ups, all of which impact the beam motion, either actively or passively. An accelerator works properly only when all of its components work precisely and cooperatively.

While accelerators are always built according to well studied design models, it is unrealistic to expect a new accelerator to realize the design performance when it is first turned on. During the commissioning period, adjustments of the machine setting have to be made to achieve the desired high performance. During the lifetime of an accelerator, when the machine configuration is intentionally modified, or if the environment conditions change, the machine setting also has to be adjusted accordingly.

The machine setting adjustments are necessary because the actual accelerator is different from the design model in many ways. Manufacturing errors, calibration errors, alignment errors, power supply fluctuation, etc., affect every component of the machine. Human errors and component malfunctions can also occur, resulting in, for example, cable swaps, reversed polarities, or short circuits. Sometimes these errors are very difficult to detect and correct. On the other hand, the design model almost always employs some simplifications in the treatment of the physical processes involved and makes some omissions of the less significant effects. The small errors on the many individual components can add up to cause substantial differences to the beam dynamics behavior between the model and reality.

Beam-based methods are essential to detect the errors in the machine and to suggest the adjustments necessary to achieve high performance. These methods can be classified into two categories: *beam-based correction* and *beam-based optimization*. The correction methods rely on the diagnostics to probe the behaviors of the machine and advanced data analysis techniques to extract useful information. The optimization methods, however, treat the machine as a black-box; they probe the parameter space by trying out new settings and use the result to guide the search for the optimal setting.

In this book we will systematically examine the two beam-based approaches for accelerators. Part I aims at providing the theoretical background for the discussion of accelerator operation challenges. Part II of the book is dedicated to beam-based correction. It covers orbit correction, linear optics measurement and correction, and linear coupling and nonlinear dynamics correction. Part III is dedicated to beam-based optimization, which includes general considerations, optimization algorithms, and examples of online optimization experiments.

Beam-based methods are an extensively researched area with a long history. This book is focused on the techniques that are deemed useful in practical accelerator operations, instead of the historical development of the methods. Although I tried to properly reference past works on the topics, inevitably some may have been inadvertently neglected. I apologize for any such cases.

I would like to take this opportunity to thank many colleagues who helped to make this work possible. My PhD advisor, Prof. S. Y. Lee, provided wise guidance in my early research that led me into the study of beam-based methods. Eric Prebys, Ray Tomlin, and Chuck Ankenbrandt supported my work at Fermilab. At SLAC, Dr. James Safranek has been a constant source of support. I benefited greatly from him through many inspiring discussions. Bob Hettel, Jim Sebek, and Jeff Corbett have also been very helpful. Special thanks go to SPEAR3 operators for their support during many accelerator physics experiments. Xi Yang of BNL provided BPM data from NSLS-II that are used in Chapters 5 and 6. Jim Sebek read Chapters 1 and 2 of the draft and provided valuable editing suggestions.

Last but not least, I would like to thank my wife, Suyan Ling, for her unwavering support on the home front.

Xiaobiao Huang
Palo Alto, CA

I

Introduction to Accelerator Physics

Basics of beam dynamics

CONTENTS

A particle beam consists of particles that move roughly with the same speed and in the same direction within a finite cross-section. In an accelerator, electromagnetic fields of various distributions in space and time are placed along the path of the beam through the accelerator components to guide the beam motion, change the beam energy, and provide focusing. Beam motion is also affected by the electromagnetic fields generated by the beam itself, directly or through the interactions with the environment. The study of beam motion under the influence of electromagnetic fields in accelerators is called beam dynamics.

It is necessary to understand the basics of beam dynamics in linacs and synchrotrons as it is the foundation for the discussions of the operation requirements and the methods of fulfilling the requirements through beam-based methods. Only single particle dynamics is covered here. The transverse dynamics describes beam motion in its deviation from the design orbit in the plane perpendicular to the design orbit. The longitudinal dynamics describes the motion in the direction along the design orbit, which involves oscillations of beam energy and arrival time.

The transverse motion of a beam in accelerators is determined by the magnetic fields in the various types of magnets. The magnets along the beam path in the accelerator constitute its lattice. Dipole magnets in the lattice determine the orbit geometry. Transverse motion is described by the deviation of particle motion from the reference orbit, which is typically the design orbit. Imperfections in the accelerator often cause the beam in an actual machine to travel on an orbit different from the reference orbit. Correction of the beam orbit toward the ideal orbit is called orbit steering or orbit correction.

Particles in a beam tend to diverge from each other as they travel along the orbit due to their slightly different directions of motion. If no intervention is taken, the transverse beam size will indefinitely grow, causing beam loss on the vacuum chamber. Quadrupole magnets provide a magnetic field that varies linearly with transverse position. Such a field bends the stray particles back toward the design orbit as particles with larger excursions receive larger correcting kicks. This is called focusing. However, a quadrupole magnet that focuses the beam in the horizontal direction necessarily defocuses it in the vertical plane and vice versa. To keep the beam focused in both transverse directions, quadrupole magnets with opposite polarities are placed alternately along the beam path. This is the alternating gradient focusing scheme, also called the strong focusing scheme. A properly designed strong focusing scheme maintains the orbit stability and keeps a compact beam size.

Steering and focusing are two basic requirements for the transverse beam motion. Magnets responsible for steering and focusing, namely dipole and quadrupole magnets, have magnetic fields that are constant or linear with the transverse position coordinates. The beam motion under such fields is linear. With a proper focusing scheme, the periodic linear motion in a circular accelerator is stable. However, synchrotrons and, in particular, storage rings, typically need sextupole magnets to correct the focusing errors for particles with energy errors (referred to as chromatic errors). The magnetic fields in sextupoles have nonlinear dependence on transverse position coordinates and hence the periodic motion becomes nonlinear. The nonlinearity causes the motion to be unstable for particles with sufficiently large offsets from the design orbit. Ensuring a large stable area is another critical requirement for the transverse motion in circular accelerators.

In this chapter we will briefly introduce the theory of transverse and longitudinal beam motion in accelerators.

1.1 BEAM MOTION IN MAGNET LATTICES

1.1.1 Hamiltonian and the equations of motion

In an accelerator the particles in a beam are expected to closely follow the design path. It is the deviations of the particles from the design path that are of our concern. Therefore, typically we adopt a moving curvilinear coordinate system to describe the particle motion, using the design orbit as the reference

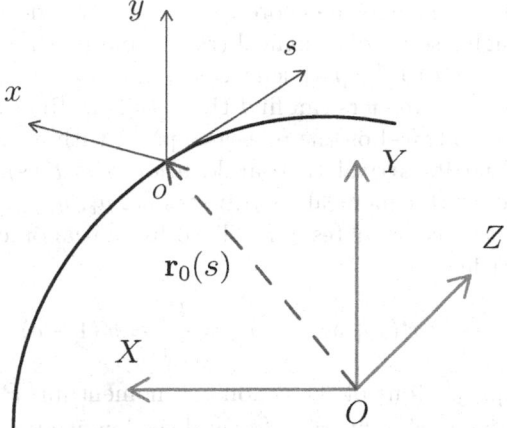

Figure 1.1 The curvilinear coordinate system for the description of beam motion.

orbit and the deviations from the reference orbit to measure the positions and directions of the particles. This coordinate system is called Frenet-Serret coordinate system, which is illustrated in Figure 1.1.

The design path is represented by a curve, $\mathbf{r}_0(s)$, in the global Cartesian coordinate system (O-XYZ), where s is the path length. At any point along the design path, the position of a particle is measured in the local coordinate system, o-xys, with the unit vectors along x, y, and s directions given as follows: $\hat{\mathbf{s}}$ is the tangent unit vector of the design path, $\hat{\mathbf{x}}$ is the normal unit vector, and $\hat{\mathbf{y}}$ is the cross product of $\hat{\mathbf{s}}$ and $\hat{\mathbf{x}}$. x and y are the two transverse directions and s is the longitudinal direction. Usually the design path is a planar curve on the horizontal plane, hence x is referred to as the horizontal direction and y the vertical direction. The design path could consist of sections that bend in the vertical direction or in a plane that is at an arbitrary angle with the horizontal plane. In such cases, the local coordinate system can rotate about the s direction at the transition points in and out of those sections.

In general, the motion of charged particles in external electromagnetic fields is governed by the Hamiltonian [68]

$$H = q\Phi + c\sqrt{m^2c^2 + (\mathbf{P} - q\mathbf{A})^2},\tag{1.1}$$

where c is the speed of light in vacuum, q and m are the charge and mass of the particle, respectively, Φ and $\mathbf{A}$ are the scalar and vector potentials of the fields, respectively, $\mathbf{P} = \mathbf{p} + q\mathbf{A}$ is the canonical momentum, and $\mathbf{p}$ is the mechanical momentum. The transverse motion is typically determined by static magnetic fields, for which $\Phi = 0$ and $\frac{\partial \mathbf{A}}{\partial t} = 0$. In principle, after the vector potential is specified over the space, the motion of a particle is completely determined by the initial conditions of the position and the canonical momentum.

It is customary to change the coordinates from the global system to the local system through a series of canonical transformations and a change of free variable from the time t to the path length s. The details of the procedure are omitted here; interested readers can find the details in Ref. [121]. A reference particle is assumed to travel on the reference path, with a constant canonical momentum, P_0, and its arrival time at location s is $t_0(s)$. The dynamical system is measured with canonical coordinates $(x, p_x, y, p_y, \Delta z, \delta)$, with the transverse momentum coordinates normalized by the canonical momentum of the reference particle

$$p_x = \frac{P_x}{P_0} \approx x'(1 + \delta), \qquad p_y = \frac{P_y}{P_0} \approx y'(1 + \delta), \tag{1.2}$$

where P_x are P_y projections of the canonical momentum, $\mathbf{P}$, on the x and y directions, respectively, $x' = \frac{dx}{ds}$, $y' = \frac{dy}{ds}$, and the longitudinal coordinates are

$$\Delta z = -\beta_0 c(t - t_0), \qquad \delta = \frac{P - P_0}{P_0}, \tag{1.3}$$

where $\beta_0 c$ is the velocity of the reference particle. Coordinate Δz is the distance between the particle and the reference particle, with $\Delta z < 0$ indicating that the particle is behind the reference particle.

The new Hamiltonian in these coordinates is

$$\begin{aligned} H = \quad & -(1 + hx)\sqrt{(1 + \delta)^2 - \frac{\delta^2}{\gamma_0^2} - (p_x - a_x)^2 - (p_y - a_y)^2} \\ & -(1 + hx)a_s + (1 + \delta), \end{aligned} \tag{1.4}$$

where $h = \frac{1}{\rho}$ is the local curvature of the reference path, ρ is the bending radius, γ_0 is the Lorentz energy factor for the reference particle, and

$$a_{xys} \equiv \frac{qA_{xys}}{P_0}, \tag{1.5}$$

are the components of the vector potential on the x, y, and s directions normalized by the magnetic rigidity of the reference particle, $B\rho = \frac{P_0}{q}$.

The Hamiltonian in Eq. (1.4) is exact, but may not be easy to solve. Since in reality the quantities $p_{x,y}$ and $a_{x,y}$ are often small, the square root in the equation can be expanded in a Taylor series, keeping only the leading terms. Small p_x and p_y correspond to the para-axial condition because $\sqrt{p_x^2 + p_y^2}$ is approximately the angle between the direction of motion and the reference path. In addition, the magnetic fields in an accelerator are typically in the transverse plane, which can be derived from vector potentials with only the A_s component, i.e., $a_x = a_y = 0$ can be assumed. Under these conditions, the Hamiltonian can be significantly simplified, to the form

$$H = (1 + hx)\left(\frac{p_x^2 + p_y^2 + \delta^2/\gamma_0^2}{2(1 + \delta)} - a_s\right) - hx(1 + \delta). \tag{1.6}$$

With the Hamiltonian in Eq. (1.6) and a vector potential $a_s(x, y, s)$ given in the local coordinate system, the beam motion can be determined.

The equations of motion for the transverse plane can be derived from the Hamiltonian, using Hamilton's equations

$$x' = \frac{\partial H}{\partial p_x}, \quad p'_x = -\frac{\partial H}{\partial x}, \quad y' = \frac{\partial H}{\partial p_y}, \quad p'_y = -\frac{\partial H}{\partial y}, \tag{1.7}$$

from which we obtain

$$x'' = -\frac{(1 + hx)^2}{1 + \delta} \frac{B_y}{B\rho} + h(1 + hx), \tag{1.8}$$

$$y'' = \frac{(1 + hx)^2}{1 + \delta} \frac{B_x}{B\rho}, \tag{1.9}$$

where $'$ and $''$ denote taking the first and second order derivatives with respect to s, respectively, and $B_{x,y}$ are magnetic field components. In the derivation we have used the formulas to calculate the magnetic fields for the curvilinear coordinate system

$$\frac{B_x}{B\rho} = \frac{\partial}{\partial y}[a_s], \quad \frac{B_y}{B\rho} = -\frac{1}{1 + hx}\frac{\partial}{\partial y}[(1 + hx)a_s], \tag{1.10}$$

which are applicable when $a_x = a_y = 0$ (see the next section).

The motion in the longitudinal plane can be derived from

$$z' = \frac{\partial H}{\partial \delta}, \quad \delta' = -\frac{\partial H}{\partial z}, \tag{1.11}$$

which give

$$z' \approx (1 + hx)\left(-\frac{x'^2 + y'^2}{2} + \frac{\delta}{\gamma_0^2}\left(1 - \frac{3}{2}\delta\right)\right) - hx, \tag{1.12}$$

and $\delta' = 0$. The momentum coordinate is a constant because the Hamiltonian Eq. (1.4) does not contain any time dependent electromagnetic fields.

1.1.2 Magnets and magnetic fields

In the current free region of a static magnetic field, the vector potential satisfies the Laplace equation, $\nabla^2 \mathbf{A} = 0$ (using the Coulomb gauge $\nabla \cdot \mathbf{A} = 0$). At locations where the reference path is a straight line (i.e., $h = 0$), the local coordinate system is Cartesian, in which case $\nabla^2 A_s = 0$. When there are only transverse magnetic fields, a solution with $A_x = A_y = 0$ can be found, for which $\frac{\partial A_s}{\partial s} = 0$ under the Coulomb gauge condition. Therefore, we have

$$\nabla_\perp^2 A_s = \frac{\partial^2 A_s}{\partial x^2} + \frac{\partial^2 A_s}{\partial y^2} = 0. \tag{1.13}$$

The solutions to Eq. (1.13) can be expanded in the form

$$A_s = -\mathrm{Re} \sum_{n=0}^{\infty} \frac{B_n + iA_n}{(n+1)!}(x+iy)^{n+1}, \tag{1.14}$$

where Re indicates taking the real part. The corresponding magnetic field can be calculated from A_s with $\mathbf{B}_\perp = \nabla_\perp \times A_s\hat{\mathbf{s}}$, which is given by

$$B_y + iB_x = \sum_{n=0}^{\infty} \frac{B_n + iA_n}{n!}(x+iy)^n, \tag{1.15}$$

where the terms corresponding to each integer, n, describe the fields for the n'th multipoles, with coefficient A_n for the skew multipole and B_n for the normal multipole. The vector potentials and magnetic fields of a few low order multipole components are listed below,

Horizontal dipole ($n = 0$):

$$A_s = -B_0x, \quad B_x = 0, \quad B_y = B_0 \tag{1.16}$$

Vertical dipole ($n = 0$):

$$A_s = A_0y, \quad B_x = A_0, \quad B_y = 0 \tag{1.17}$$

Normal quadrupole ($n = 1$):

$$A_s = -B_1\frac{x^2 - y^2}{2}, \quad B_x = B_1y, \quad B_y = B_1x \tag{1.18}$$

Skew quadrupole ($n = 1$):

$$A_s = A_1xy, \quad B_x = A_1x, \quad B_y = -A_1y \tag{1.19}$$

Normal sextupole ($n = 2$):

$$A_s = -B_2\frac{x^3 - 3xy^2}{6}, \quad B_x = B_2xy, \quad B_y = B_2\frac{x^2 - y^2}{2} \tag{1.20}$$

Skew sextupole ($n = 2$):

$$A_s = A_2\frac{3x^2y - y^3}{6}, \quad B_x = A_2\frac{x^2 - y^2}{2}, \quad B_y = -A_2xy \tag{1.21}$$

The multipole expansion in Eqs. (1.14-1.15) is valid for magnets with a straight geometry ($h = 0$), including multipoles with $n \geq 1$ (quadrupoles and higher order multipoles) and orbit corrector magnets (weak dipole magnets that do not change the design path). However, the straight geometry condition

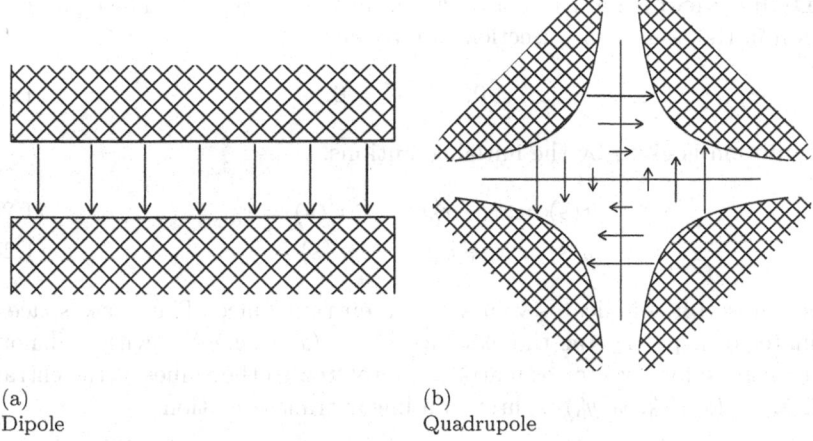

(a)
Dipole

(b)
Quadrupole

Figure 1.2   Magnetic fields in dipole and quadrupole magnets.

does not apply to dipole magnets in which the reference orbit is curved. With nonzero curvature ($|h| > 0$), the multipole expansion can be derived from the Laplace equation in the curvilinear system. If there are only vertical fields on the midplane ($y = 0$), which is given by

$$B_y(x) = B_0 + B_1 x + B_2 \frac{x^2}{2!} + \cdots , \qquad (1.22)$$

the vector potential A_s takes the form [67]

$$
\begin{aligned}
A_s \;=\; & -B_0 \left(x - \frac{hx^2}{2(1 + hx)} \right) - B_1 \left(\frac{1}{2}(x^2 - y^2) - \frac{h}{6}x^3 + \cdots \right) \\
& -B_2 \left(\frac{1}{6}(x^3 - 3x^2 y) + \cdots \right) + \cdots ,
\end{aligned}
\qquad (1.23)
$$

and $A_x = A_y = 0$. A pure dipole consists of only the B_0 term, while a combined-function dipole magnet has both B_0 and B_1 terms.

The magnetic field distributions in the transverse plane in dipole and quadrupole magnets are illustrated in Figure 1.2.

1.2   TRANSVERSE DYNAMICS

1.2.1   Beam motion in linear components

Knowing the magnetic fields, the particle motion through an accelerator component can be solved from Eqs. (1.8-1.9) and Eq. (1.12). The equations of motion in three types of components - drift space, dipole magnet, and quadrupole - are linear and hence can be readily solved. The motion in these components determines the linear optics of the accelerator beam line.

Drift space: In a drift space, $h = 0$ and $B_x = B_y = 0$. The equations of motion in the transverse directions are reduced to

$$x'' = 0, \qquad y'' = 0. \tag{1.24}$$

The solution is given by the initial conditions,

$$x(s) = x_0 + x_0's, \qquad x'(s) = x_0', \tag{1.25a}$$
$$y(s) = y_0 + y_0's, \qquad y'(s) = y_0', \tag{1.25b}$$

where subscript 0 indicates values at the entrance face. The phase space coordinates of a particle at the exit face, $\mathbf{X} = (x, x', y, y')^T$ (with ^{T} denoting the transpose for a vector or matrix), are related to the values at the entrance face, $\mathbf{X}_0 = (x_0, x_0', y_0, y_0')^T$, through a linear transformation

$$\mathbf{X} = \mathbf{M}\mathbf{X}_0, \tag{1.26}$$

where matrix $\mathbf{M}$ is referred to as the transfer matrix for the element. For the drift space with length L, the transfer matrix is

$$\mathbf{M}_{\text{drift}} = \begin{pmatrix} 1 & L & 0 & 0 \\ 0 & 1 & 0 & 0 \\ 0 & 0 & 1 & L \\ 0 & 0 & 0 & 1 \end{pmatrix}. \tag{1.27}$$

In a drift space, the particle simply moves along a straight line specified by its initial direction. Particles with different initial angle coordinates will diverge in the position coordinates as they travel in a drift space.

Dipole magnet: In a dipole magnet, the equations of motion can be derived from the Hamiltonian Eq. (1.6) and the vector potential Eq. (1.23). Typically the reference orbit in the dipole magnet is the circular orbit defined by the dipole component, B_0, for which the curvature satisfies $h = b_0 \equiv \frac{B_0}{B\rho}$. In the case of a combined-function dipole magnet with quadrupole component B_1, the equations of motion to the linear order of the coordinates are

$$x'' + (b_1 + h^2)x = h\delta, \qquad y'' - b_1 y = 0, \tag{1.28}$$

where $b_1 \equiv \frac{B_1}{B\rho}$ is the normalized gradient. Labeling $K_x = b_1 + h^2$ and $K_y = -b_1$, the solution to Eq. (1.28) can be written in the matrix form as

$$\mathbf{X}(s) = \mathbf{M}(s|0)\mathbf{X}_0 + \delta\mathbf{d}(s), \tag{1.29}$$

where $\mathbf{M}(s|0)$ is the transfer matrix from the entrance point ($s = 0$) to point s, and $\delta\mathbf{d}(s)$ is the particular solution that accounts for the inhomogeneous term $h\delta$ in Eq. (1.28). The term $\delta\mathbf{d}(s)$ represents the trajectory deviations for off-energy particles, which give rise to dispersion.

For a pure dipole, $b_1 = 0$, and the solution for the transfer matrix is

$$\mathbf{M}(s|0) = \begin{pmatrix} \cos\theta & \rho\sin\theta & 0 & 0 \\ -\frac{\sin\theta}{\rho} & \cos\theta & 0 & 0 \\ 0 & 0 & 1 & \rho\theta \\ 0 & 0 & 0 & 1 \end{pmatrix}, \tag{1.30}$$

$$\mathbf{d}(s) = \begin{pmatrix} \rho(1-\cos\theta) & \sin\theta & 0 & 0 \end{pmatrix}^T, \tag{1.31}$$

where $\theta = hs$, and the bending radius is $\rho = 1/h$. The pure dipole magnet provides horizontal focusing and behaves like a drift space in the vertical plane.

For the case with $K_x > 0$ and $K_y < 0$, the transfer matrix and the particular solution are given by

$$\mathbf{M}(s|0) = \begin{pmatrix} \cos k_x s & \frac{\sin k_x s}{k_x} & 0 & 0 \\ -k_x \sin k_x s & \cos k_x s & 0 & 0 \\ 0 & 0 & \cosh k_y s & \frac{\sinh k_y s}{k_y} \\ 0 & 0 & k_y \sinh k_y s & \cosh k_y s \end{pmatrix}, \tag{1.32}$$

$$\mathbf{d}(s) = \begin{pmatrix} \frac{1-\cos k_x s}{k_x^2 \rho} & \frac{\sin k_x s}{k_x \rho} & 0 & 0 \end{pmatrix}^T, \tag{1.33}$$

where $k_x = \sqrt{K_x}$ and $k_y = \sqrt{-K_y}$. In the above case, the magnet focuses in the horizontal plane and defocuses in the vertical plane. The solution for the case with $K_x < 0$ and $K_y > 0$ is similar; the magnet now defocuses in the horizontal plane and focuses in the vertical plane.

In the case $-h^2 < b_1 < 0$, both K_x and K_y are positive; hence the dipole magnet provides focusing for both transverse planes. It is customary to define the focusing index $n = -b_1\rho^2$. Then the equations of motion become

$$x'' + (1-n)h^2 x = 0, \qquad y'' + nh^2 y = 0. \tag{1.34}$$

This is the case of weak focusing. Because the transverse beam size in weak-focusing accelerators tends to be very large, modern accelerators usually employ the strong focusing scheme instead.

The magnetic field in the transition region at the edges of a dipole magnet can provide additional focusing or defocusing if the beam orbit is not perpendicular to the magnet face. If the reference orbit enters the dipole magnet with an angle, δ_1, with respect to the normal of the entrance face of the magnet, as illustrated in Figure 1.3, a particle that comes to the entrance point of the reference orbit with a horizontal offset x will see more bending (than the reference particle) if $x < 0$ and less bending if $x > 0$. Therefore the entrance edge gives the beam defocusing in the horizontal plane.

The magnetic field has a component, $B_Z \propto y$, in the direction normal to the entrance face in the transition area in which the bending field B_y goes from zero to full strength, as is required by the condition $\frac{\partial B_Z}{\partial Y} - \frac{\partial B_Y}{\partial Z} = 0$. This B_Z field has a projection onto the x direction, which gives the beam vertical

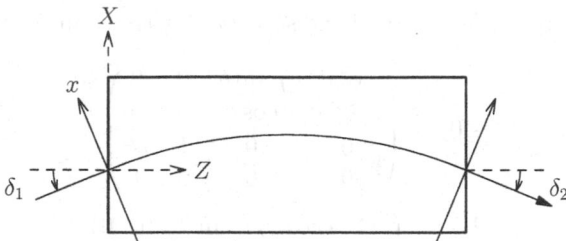

Figure 1.3 Edge focusing at the entrance and exit faces of a dipole magnet. The y and Y directions point out of the paper.

focusing. The edge focusing effect at the entrance face is represented by the transfer matrix

$$\mathbf{X}_+ = \mathbf{M}\mathbf{X}_-, \qquad \mathbf{M} = \begin{pmatrix} 1 & 0 & 0 & 0 \\ h\tan\delta_1 & 1 & 0 & 0 \\ 0 & 0 & 1 & 0 \\ 0 & 0 & -h\tan\delta_1 & 1 \end{pmatrix}, \tag{1.35}$$

where the subscripts $+$ and $-$ stand for the coordinates before and after the edge, respectively.

Edge focusing at the exit face is similar. The transfer matrices are the same as in Eq. (1.35), except with δ_2 substituted for δ_1, when the definition of the entrance and exit angles follow the convention in Figure 1.3.

Quadrupole magnet: Including the effect of energy errors, the equations of motion in a quadrupole magnet are

$$x'' + \frac{b_1}{1+\delta}x = 0, \qquad y'' - \frac{b_1}{1+\delta}y = 0. \tag{1.36}$$

Defining $K = \frac{b_1}{1+\delta}$, the solution for the case $K > 0$ is represented by the transfer matrix

$$\mathbf{M}_{FQ}(s) = \begin{pmatrix} \cos\sqrt{K}s & \frac{\sin\sqrt{K}s}{\sqrt{K}} & 0 & 0 \\ -\sqrt{K}\sin\sqrt{K}s & \cos\sqrt{K}s & 0 & 0 \\ 0 & 0 & \cosh\sqrt{K}s & \frac{\sinh\sqrt{K}s}{\sqrt{K}} \\ 0 & 0 & \sqrt{K}\sinh\sqrt{K}s & \cosh\sqrt{K}s \end{pmatrix}. \tag{1.37}$$

The transfer matrix for the case with $K < 0$ can be obtained by swapping the 2-by-2 blocks for the horizontal and vertical planes and replacing K with $|K|$.

Typically for a quadrupole magnet $\sqrt{|K|}L$ is much less than unity. For example, in the SPEAR3 storage ring, the typical value of $\sqrt{|K|}L$ for QF magnets is about 0.45. For the purpose of a rough estimate, the transfer matrix is often approximated with the thin-lens limit, by taking $\sqrt{|K|}L \to 0$

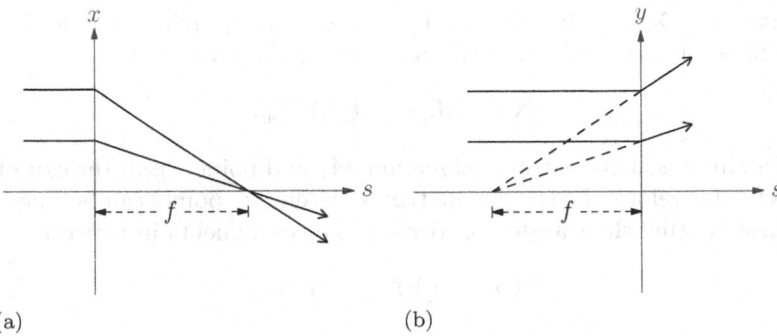

(a)   (b)

Figure 1.4 Illustration of a thin-lens focusing quadrupole which (a) focuses in the horizontal plane (b) and defocuses in the vertical plane.

while keeping KL constant. The quadrupole transfer matrix for $K > 0$ in the thin-lens limit is

$$\mathbf{M}(K > 0) = \begin{pmatrix} 1 & 0 & 0 & 0 \\ -\frac{1}{f} & 1 & 0 & 0 \\ 0 & 0 & 1 & 0 \\ 0 & 0 & \frac{1}{f} & 1 \end{pmatrix}, \tag{1.38}$$

with the focal length $f = \frac{1}{KL}$. Across a thin-lens quadrupole, the position coordinates do not change, but the angle coordinates will change according to the initial position coordinates. For the $K > 0$ case,

$$\Delta x' = -\frac{x}{f}, \qquad \Delta y' = \frac{y}{f}, \tag{1.39}$$

which indicates that the quadrupole focuses the beam in the horizontal plane and defocuses the beam in the vertical plane. This is the case illustrated in Figure 1.4. By convention, a quadrupole that focuses in the horizontal plane is called a focusing quadrupole. Conversely, when $K < 0$, the quadrupole defocuses on the horizontal plane and is called a defocusing quadrupole.

1.2.2 Transfer matrix and transfer map

Drift spaces, dipole magnets, and quadrupole magnets are the basic building blocks of accelerator lattices. The placement of these components in a beam line determines the transformation of the phase space coordinates to the linear order. The properties of such linear transformations are called the linear optics of the beam line. Transfer matrices are a basic representation of linear optics.

The transfer matrix between two points in the lattice, points 0 and n, relates the phase space coordinate vectors at the two points through the equation

$$\mathbf{X}_n = \mathbf{M}(n|0)\mathbf{X}_0. \tag{1.40}$$

Alternatively, $\mathbf{X}_n$ can be obtained by successively applying the transfer matrices of accelerator sections between the two points, i.e.,

$$\mathbf{X}_n = \mathbf{M}_n \cdots \mathbf{M}_2 \mathbf{M}_1 \mathbf{X}_0, \tag{1.41}$$

where point 0 is at the entrance of section $\mathbf{M}_1$ and point n is at the exit of section $\mathbf{M}_n$. Therefore, the transfer matrix between two points can be calculated by concatenating the transfer matrices of the components in between,

$$\mathbf{M}(n|0) = \mathbf{M}_n \cdots \mathbf{M}_2 \mathbf{M}_1. \tag{1.42}$$

The transfer matrix is the Jacobian matrix of the canonical transformation that relates the phase space coordinate vectors at the two points, i.e.,

$$\mathbf{M}(n|0)_{ij} = \frac{\partial (\mathbf{X}_n)_i}{\partial (\mathbf{X}_0)_j}. \tag{1.43}$$

It is well known that the Jacobian matrix of a canonical transformation is symplectic, which means it satisfies the condition

$$\mathbf{M}^T \mathbf{S} \mathbf{M} = \mathbf{S}, \tag{1.44}$$

with the anti-symmetric matrix for the case of 2-dimensional motion given by

$$\mathbf{S} = \begin{pmatrix} \mathbf{S}_2 & \mathbf{0} \\ \mathbf{0} & \mathbf{S}_2 \end{pmatrix}, \quad \mathbf{S}_2 = \begin{pmatrix} 0 & 1 \\ -1 & 0 \end{pmatrix}. \tag{1.45}$$

The symplectic condition, Eq. (1.44), requires the determinant of the transfer matrix to be unity,

$$\det \mathbf{M} = 1. \tag{1.46}$$

This indicates that the volume enclosed by a surface in the phase space will be preserved as the surface evolves according to the Hamiltonian.

For a full description of the transverse motion, 4×4 transfer matrices are used for the transformation of coordinates (x, x', y, y'). However, when there is no coupling between the horizontal and vertical planes, the 2×2 off-diagonal blocks of the transfer matrix are zeros. In this case, the horizontal and vertical motion are decoupled and can be described separately, with the top and bottom 2×2 diagonal blocks, respectively. These transfer matrices also satisfy the symplectic condition. It can be shown that a 2×2 matrix is symplectic if and only if its determinant is unity.

The transfer matrix does not describe the effects of the nonlinear fields in the accelerator elements. Including the nonlinear motion, the general effect of an element or an accelerator section can be represented by a map between the phase space coordinates at the entrance and exit faces

$$\mathbf{X}_f = \mathbf{M}(\mathbf{X}_i) \equiv \mathbf{M} \mathbf{X}_i, \tag{1.47}$$

where $\mathbf{MX}_i$ denotes the full map, not just the transfer matrix. The map can be expanded into a Taylor series

$$\mathbf{X}_f = \Delta\mathbf{X}_0 + \mathbf{R}\mathbf{X}_i + \mathbf{X}_i^T\mathbf{T}\mathbf{X}_i + \cdots, \tag{1.48}$$

where $\Delta\mathbf{X}_0$ is a constant coordinate shift that represents the accumulated effects of dipole kicks throughout the section, $\mathbf{R}$ is the linear transfer matrix, and $\mathbf{T}$ is the second order map. It is conventional to denote the Taylor map up to the second order the TRANSPORT map [17].

The map of a composite section can be obtained by concatenating the map of the individual elements sequentially.

1.2.3 Strong focusing principle and orbit stability

As pointed out previously, a quadrupole magnet always focuses in one transverse plane and defocuses in the other. This is unlike a convex lens for light optics, which focuses simultaneously in both transverse planes. Therefore, according to the strong focusing principle, focusing and defocusing quadrupoles are alternately placed along the beam path to keep the particle beam focused in both planes.

Alternate focusing is often implemented with repetitive, identical magnet lattice cells. The FODO cell is a basic cell structure, which consists of one focusing quadrupole and one defocusing quadrupole, separated by a drift space, as illustrated in Figure 1.5. The transfer matrix for the cell, starting from the center of the focusing quadrupole (QF) to the center of the next QF, can be calculated using the transfer matrices of the individual components

$$\mathbf{M}_F = \mathbf{M}_{\frac{1}{2}QF}\mathbf{M}_D\mathbf{M}_{QD}\mathbf{M}_D\mathbf{M}_{\frac{1}{2}QF}, \tag{1.49}$$

where subscript "D" stands for the drift space. Using the thin-lens approximation for the quadrupoles and assuming the defocusing quadrupole (QD) and the QF have the same focal length, the horizontal transfer matrix for the FODO cell at the QF center is found to be

$$\begin{aligned}
\mathbf{M}_{x,F} &= \begin{pmatrix} 1 & 0 \\ -\frac{1}{2f} & 1 \end{pmatrix} \begin{pmatrix} 1 & L \\ 0 & 1 \end{pmatrix} \begin{pmatrix} 1 & 0 \\ \frac{1}{f} & 1 \end{pmatrix} \begin{pmatrix} 1 & L \\ 0 & 1 \end{pmatrix} \begin{pmatrix} 1 & 0 \\ -\frac{1}{2f} & 1 \end{pmatrix} \\
&= \begin{pmatrix} 1 - \frac{L^2}{2f^2} & 2L\left(1 + \frac{L}{2f}\right) \\ -\frac{L}{2f^2}\left(1 - \frac{L}{2f}\right) & 1 - \frac{L^2}{2f^2} \end{pmatrix},
\end{aligned} \tag{1.50}$$

where L is half the length of the cell and f is the focal length.

The transfer matrix for the vertical plane at the QF center can be obtained from Eq. (1.50) by reversing the sign of f. It can be seen that if $2f$ is considerably larger than L, the FODO cell provides focusing in both transverse planes since the (2, 1) elements of the horizontal and vertical transfer matrices will both be negative, while the diagonal elements remain positive.

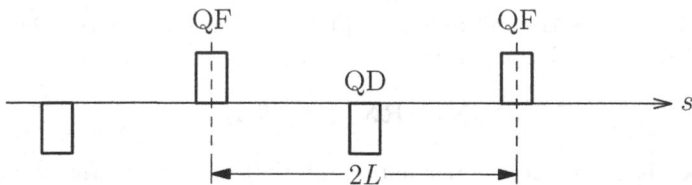

Figure 1.5 Illustration of the strong focusing principle with FODO cells.

The stability of beam motion is an important requirement for the design of periodic lattices. For a stable lattice, a particle launched in the vicinity of the phase space origin (representing the reference orbit) will stay around the origin after traveling through many periods. The orbit stability of a periodic linear lattice can be analyzed through the transfer matrix of one periodic cell. Considering the motion in one plane, the cell transfer matrix can be transformed to the form

$$\mathbf{M} = \mathbf{V}\mathbf{\Lambda}\mathbf{V}^{-1}, \qquad \mathbf{\Lambda} = \mathrm{diag}(\lambda, \frac{1}{\lambda}), \tag{1.51}$$

where, as a consequence of the symplecticity of $\mathbf{M}$, λ and $\frac{1}{\lambda}$ are both eigenvalues of $\mathbf{M}$, $\mathbf{\Lambda}$ is a diagonal matrix with diagonal elements λ and $\frac{1}{\lambda}$, and columns in $\mathbf{V}$ are the corresponding eigenvectors. For a particle with initial coordinates $\mathbf{X}_0$, the coordinates after m cells will be

$$\mathbf{X}_m = \mathbf{M}^m \mathbf{X}_0 = \mathbf{V}\mathbf{\Lambda}^m \mathbf{V}^{-1} \mathbf{X}_0. \tag{1.52}$$

Because $\mathbf{\Lambda}^m = \mathrm{diag}(\lambda^m, \frac{1}{\lambda^m})$, the particle motion in the lattice is stable if and only if $|\lambda| = 1$.

The eigenvalues of matrix $\mathbf{M}$ can be found by solving the equation $\det(\mathbf{M} - \lambda \mathbf{I}) = 0$. For the case of one-dimensional motion, this becomes

$$\lambda^2 - \mathrm{Tr}(\mathbf{M})\lambda + 1 = 0, \tag{1.53}$$

where $\mathrm{Tr}(\mathbf{M}) = M_{11} + M_{22}$ is the trace of the 2×2 matrix, and we have used $\det(\mathbf{M}) = 1$. The eigenvalues are

$$\lambda_{1,2} = \frac{1}{2}\left(\mathrm{Tr}(\mathbf{M}) \pm \sqrt{\mathrm{Tr}(\mathbf{M})^2 - 4}\right), \tag{1.54}$$

which, combined with the $|\lambda| = 1$ requirement, leads to the stability condition

$$|\mathrm{Tr}(\mathbf{M})| \le 2, \tag{1.55}$$

in which case the eigenvalues of the transfer matrix are a pair of complex conjugates, $\lambda_1 = e^{i\Phi}$ and $\lambda_2 = e^{-i\Phi}$, where Φ is a real angle satisfying

$$\cos \Phi = \frac{1}{2}\mathrm{Tr}(\mathbf{M}). \tag{1.56}$$

Applying the stability condition to the FODO cell of Eq. (1.50), we find that the beam motion is stable if and only if

$$f \geq \frac{L}{2}. \tag{1.57}$$

1.2.4 Courant-Snyder parametrization

The 2×2 transfer matrix of a stable periodic lattice cell can be parametrized as follows [26]

$$\mathbf{M} = \begin{pmatrix} \cos \Phi + \alpha \sin \Phi & \beta \sin \Phi \\ -\gamma \sin \Phi & \cos \Phi - \alpha \sin \Phi \end{pmatrix}, \tag{1.58}$$

where $\gamma = \frac{1+\alpha^2}{\beta}$ and Φ is defined as the betatron phase advance over the period. Parameters α, β, and γ are called Courant-Snyder (C-S) parameters. The C-S parameters are the main characteristics of the linear optics, which are often called the linear optics functions. The β parameter is referred to as the beta function. The cell transfer matrix can be further rewritten as

$$\mathbf{M} = \mathbf{B}\mathbf{R}(\Phi)\mathbf{B}^{-1}, \tag{1.59}$$

with

$$\mathbf{B} = \begin{pmatrix} \sqrt{\beta} & 0 \\ -\frac{\alpha}{\sqrt{\beta}} & \frac{1}{\sqrt{\beta}} \end{pmatrix}, \quad \text{and} \quad \mathbf{R}(\psi) = \begin{pmatrix} \cos \psi & \sin \psi \\ -\sin \psi & \cos \psi \end{pmatrix}. \tag{1.60}$$

The transverse coordinate of a particle oscillates in the lattice. This is called the betatron oscillation. The betatron phase advance for a full revolution of a circular accelerator, Φ, is used to define the betatron tune,

$$\nu = \frac{\Phi}{2\pi}. \tag{1.61}$$

The tune is the number of oscillations the beam executes in one revolution.

If we observe the motion of a particle traveling through a periodic lattice in each cell, the coordinates observed at $s_n = s_0 + nL$, will be related to the initial coordinate, $\mathbf{X}_0$, via

$$\mathbf{X}_n = \mathbf{M}^n \mathbf{X}_0 = \mathbf{B}\mathbf{R}(n\Phi)\mathbf{B}^{-1}\mathbf{X}_0, \tag{1.62}$$

where s_0 is the initial location and L is the cell length. In the (y, y') phase space the points representing the successive observed coordinates will trace out an ellipse

$$y^2 + (\alpha y + \beta y')^2 = 2\beta J, \tag{1.63}$$

where J is an invariant of motion given by the initial coordinates. The shape and orientation of the ellipse are determined by the C-S parameters.

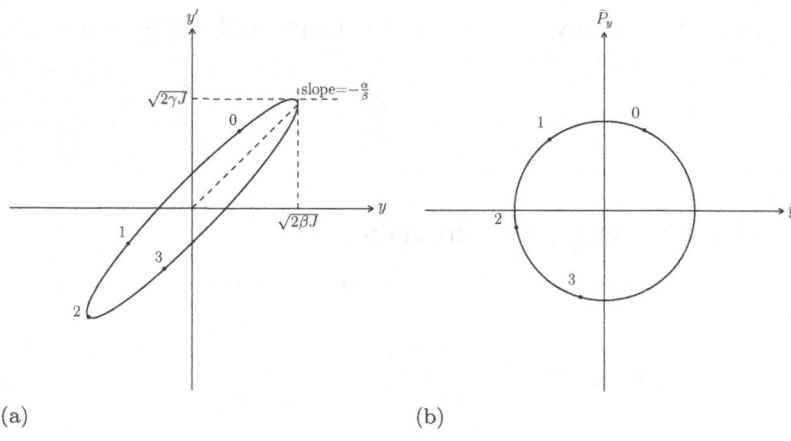

Figure 1.6 Phase space traces of a particle traveling in periodic cells. (a) in (y, y') coordinates; (b) in normalized coordinates $(\bar{y}, \bar{P}_y)$.

Eq. (1.62) indicates that we can define the normalized coordinates, $(\bar{y}, \bar{P}_y)$,

$$\bar{\mathbf{X}} \equiv \begin{pmatrix} \bar{y} \\ \bar{P}_y \end{pmatrix} = \mathbf{B}^{-1}\mathbf{X} = \frac{1}{\sqrt{\beta}} \begin{pmatrix} y \\ \alpha y + \beta y' \end{pmatrix}. \qquad (1.64)$$

The cell transfer matrix in the new coordinates represents a simple rotation through the angle Φ. In the normalized coordinates, the particle traces out a circle of radius $\sqrt{2J}$ in the phase space. The motion of a particle through periodic cells observed at a fixed location in each cell is illustrated in Figure 1.6 in both the (y, y') coordinates and $(\bar{y}, \bar{P}_y)$ coordinates.

Applying the Courant-Snyder parametrization to the FODO cell in Eq. (1.50), the phase advances is found to be

$$\Phi_{x,y} = \Phi = \sin^{-1}\frac{L}{2f}, \qquad (1.65)$$

and the beta functions at the quadrupole centers are

$$\beta_{x,\mathrm{QF}} = \beta_{y,\mathrm{QD}} = 2L\frac{1 + \sin\frac{\Phi}{2}}{\sin\Phi}, \qquad (1.66a)$$

$$\beta_{y,\mathrm{QF}} = \beta_{x,\mathrm{QD}} = 2L\frac{1 - \sin\frac{\Phi}{2}}{\sin\Phi}. \qquad (1.66b)$$

In the horizontal plane, the maximum beta function is at the QF and the minimum beta function is at the QD; the opposite is true for the vertical plane.

1.2.5 Propagation of linear optics functions

The cell transfer matrix depends on the location. The cell transfer matrices at two locations are connected through the transfer matrix between the two

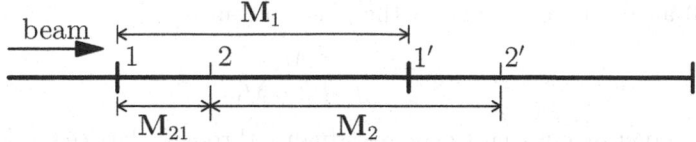

Figure 1.7 Connection of cell transfer matrices at two locations.

locations. Suppose the cell transfer matrices at points 1 and 2 are $\mathbf{M}_1$ and $\mathbf{M}_2$, respectively, and the transfer matrix from point 1 to point 2 is $\mathbf{M}_{21}$, as illustrated in Figure 1.7. The coordinate vectors at points 1 and 2 and one cell length downstream are related by

$$\mathbf{X}_{1'} = \mathbf{M}_1 \mathbf{X}_1, \qquad \mathbf{X}_{2'} = \mathbf{M}_2 \mathbf{X}_2,$$
$$\mathbf{X}_2 = \mathbf{M}_{21} \mathbf{X}_1, \qquad \mathbf{X}_{2'} = \mathbf{M}_{21} \mathbf{X}_{1'}.$$

Therefore

$$\mathbf{X}_{2'} = \mathbf{M}_2 \mathbf{X}_2 = \mathbf{M}_2 \mathbf{M}_{21} \mathbf{X}_1, \quad \mathbf{X}_{2'} = \mathbf{M}_{21} \mathbf{X}_{1'} = \mathbf{M}_{21} \mathbf{M}_1 \mathbf{X}_1.$$

Comparing the right hand side of the two equations, we obtain

$$\mathbf{M}_2 \mathbf{M}_{21} = \mathbf{M}_{21} \mathbf{M}_1,$$

and in turn

$$\mathbf{M}_2 = \mathbf{M}_{21} \mathbf{M}_1 \mathbf{M}_{21}^{-1}. \tag{1.67}$$

Eq. (1.67) means $\mathbf{M}_2$ and $\mathbf{M}_1$ are related through a similarity transformation. Therefore, their eigenvalues are the same and, consequently, the phase advances for a cell are the same measured from any location.

However, the C-S parameters vary with location. The transfer matrix from point 1 to point 2 can be written in the form

$$\mathbf{M}_{21} = \mathbf{B}_2 \mathbf{R}(\psi_{21}) \mathbf{B}_1^{-1}, \tag{1.68}$$

where $\mathbf{B}_{1,2}$ are the $\mathbf{B}$ matrices defined in Eq. (1.60) using the C-S parameters for the two locations and $\mathbf{R}$ is the rotation matrix with the rotation angle ψ_{21}, the phase advance from point 1 to point 2. Using Eq. (1.68), the relationship between the C-S parameters at the two locations is found to be

$$\mathbf{B}_2 \mathbf{B}_2^T = \mathbf{M}_{21} \mathbf{B}_1 \mathbf{B}_1^T \mathbf{M}_{21}^T. \tag{1.69}$$

Explicitly, the relations are given in terms of the elements of $\mathbf{M}_{21}$ by

$$\begin{pmatrix} \beta_2 \\ \alpha_2 \\ \gamma_2 \end{pmatrix} = \begin{pmatrix} M_{11}^2 & -2M_{11}M_{12} & M_{12}^2 \\ -M_{11}M_{21} & M_{11}M_{22} + M_{12}M_{21} & -M_{12}M_{22} \\ M_{21}^2 & -2M_{21}M_{22} & M_{22}^2 \end{pmatrix} \begin{pmatrix} \beta_1 \\ \alpha_1 \\ \gamma_1 \end{pmatrix}. \tag{1.70}$$

Eq. (1.68) also gives a formula for the phase advance

$$\psi_{21} = \tan^{-1} \frac{M_{12}}{M_{11}\beta_1 - M_{12}\alpha_1}. \tag{1.71}$$

It is worth pointing out that the phase advance through a lattice section is not uniquely determined by the transfer matrix of the section itself. It requires that the C-S parameters be given at one location on the section (such as the entrance point). In a ring lattice, the C-S parameters are naturally defined through the periodic condition. However, in a one-pass system such as a linac or a transport line, especially short ones that lack periodicity, the definition of the phase advance through the line is somewhat arbitrary.

Eqs. (1.70-1.71) can be applied to the case when the section between points 1 and 2 is an infinitesimal slice of a general linear lattice element. If the length is ds and the focusing gradient is K, the transfer matrix from point 1 to 2 is (see Eq. (1.37))

$$\mathbf{M}_{21} = \begin{pmatrix} 1 & ds \\ -Kds & 1 \end{pmatrix} + O(ds^2), \tag{1.72}$$

where $O(ds^2)$ stands for terms of ds^2 or higher. From Eq. (1.70) we obtain

$$\beta' = -2\alpha, \quad \alpha' = K\beta - \gamma, \quad \gamma' = 2\alpha K, \tag{1.73}$$

where $'$ stands for taking the derivative $\frac{d}{ds}$. The first two equations lead to a differential equation for the beta function

$$\frac{\beta''}{2} + \beta K(s) - \frac{1}{\beta}\left(\frac{1}{4}\beta'^2 + 1\right) = 0. \tag{1.74}$$

If the focusing function, $K(s)$, is given on the cell, the beta function can be calculated by solving for the periodic solution of Eq. (1.74).

Eqs. (1.71) and (1.72) give another useful result,

$$\frac{d\psi}{ds} = \frac{1}{\beta}, \tag{1.75}$$

which states that the rate of phase advance accumulation through the lattice is inversely proportional to the beta function.

Applying Eqs. (1.70-1.71) to a lattice element, we can find out how the C-S parameters change within or across the element. For example, in a drift space, we have

$$\beta(s) = \beta_0 - 2\alpha_0 s + \gamma_0 s^2, \quad \alpha(s) = \alpha_0 - \gamma_0 s, \quad \gamma(s) = \gamma_0, \tag{1.76}$$

where the subscript 0 indicates values at the point with $s = 0$. If $s = 0$ is a "waist", a symmetry point where $\alpha_0 = 0$ and $\beta_0 = \beta^*$, the beta function and phase advance are given by

$$\beta(s) = \beta^* + \frac{s^2}{\beta^*}, \quad \psi(s) = \tan^{-1}\frac{s}{\beta^*}. \tag{1.77}$$

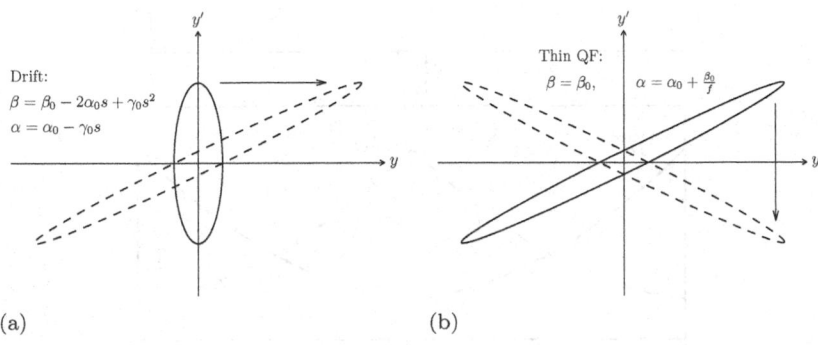

Figure 1.8 Evolution of the phase space ellipse in (a) a drift space and (b) a thin focusing quadrupole.

Across a focusing ($K > 0$) thin quadrupole, the C-S parameters are related through

$$\beta_2 = \beta_1, \quad \alpha_2 = \alpha_1 + \frac{\beta_1}{f}, \quad \gamma_2 = \gamma_1 + 2\frac{\alpha_1}{f} + \frac{\beta_1}{f^2}, \tag{1.78}$$

where the subscript 1 indicates the entrance face, 2 the exit face, and $f = \frac{1}{|K|L}$ is the focal length. For a defocusing thin quadrupole, simply reverse the sign of f in the above equation. The phase advance does not change across a thin quadrupole.

The changes of C-S parameters in the elements correspond to changes of the phase space ellipse. Figure 1.8 shows the changes across a drift space and a thin focusing quadrupole. In a drift space, the angle coordinate does not change while the position coordinate shifts linearly with y'. Therefore the ellipse is sheared along the position axis. Across a thin quadrupole the situation is the opposite: the position coordinate does not change while the angle coordinate shifts linearly with y and hence the ellipse is tilted in the y' direction.

In an accelerator with strong focusing, quadrupoles and drift spaces play dominant roles in laying out the linear optics. Knowing how the phase space ellipse and the optics functions change in these two types of elements is very useful for understanding the linear optics in accelerators. As an example, the beta functions in a FODO cell are shown in Figure 1.9.

1.2.6 Beam distribution

The linear optics of an accelerator beam line is closely related to the evolution of the transverse beam distribution. The beam distribution in the phase space (y, y') can be described by the probability density function $\rho(y, y')$, which can be characterized by its first and second order moments. The first order

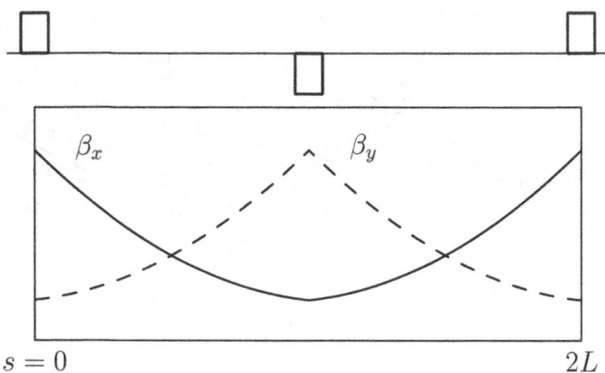

Figure 1.9   Beta functions in a FODO cell.

moments are the center of the distribution defined as

$$\langle y \rangle = \int y \rho(y, y') dy dy', \qquad \langle y' \rangle = \int y' \rho(y, y') dy dy'. \qquad (1.79)$$

We assume a distribution centered on the reference orbit, i.e., with $\langle y \rangle = \langle y' \rangle = 0$. The second-order moments are given by

$$\Sigma \equiv \begin{pmatrix} \sigma_y^2 & \sigma_{yy'} \\ \sigma_{yy'} & \sigma_{y'}^2 \end{pmatrix} = \int \rho(y, y') dy dy' \begin{pmatrix} y^2 & yy' \\ yy' & y'^2 \end{pmatrix}. \qquad (1.80)$$

The second order moment matrix can be parametrized as follows,

$$\begin{pmatrix} \sigma_y^2 & \sigma_{yy'} \\ \sigma_{yy'} & \sigma_{y'}^2 \end{pmatrix} = \epsilon_{\text{rms}} \begin{pmatrix} \bar{\beta} & -\bar{\alpha} \\ -\bar{\alpha} & \bar{\gamma} \end{pmatrix}, \qquad (1.81)$$

where the rms emittance is defined as

$$\epsilon_{\text{rms}} \equiv \sqrt{\det(\Sigma)}, \qquad (1.82)$$

and hence $\bar{\beta}\bar{\gamma} = 1 + \bar{\alpha}^2$ is required. The second order moment matrix defines an ellipse in the phase space, $\mathbf{Y}^T \Sigma^{-1} \mathbf{Y} = 1$, with $\mathbf{Y} = (y, y')^T$, which can be written as

$$y^2 + (\bar{\alpha}y + \bar{\beta}y')^2 = \bar{\beta}\epsilon_{\text{rms}}. \qquad (1.83)$$

The area of the ellipse is the rms emittance. The $\bar{\beta}$, $\bar{\alpha}$, and $\bar{\gamma}$ parameters represent the shape and orientation of the ellipse in the same manner as β, α, and γ for the phase space ellipse in Figure 1.6. If the beam distribution is Gaussian, the probability density function is given by $\rho(y, y') = \frac{1}{2\pi\epsilon_{\text{rms}}} \exp(-\frac{1}{2}\mathbf{Y}^T\Sigma^{-1}\mathbf{Y})$. In this case, an enlarged ellipse with an area of $6\sigma_{\text{rms}}$ will cover 95% of the particles in the beam [78].

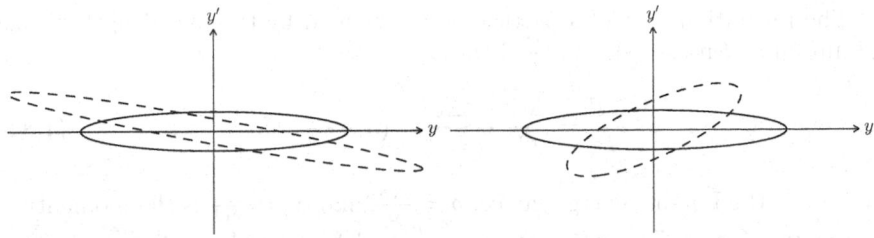

Figure 1.10 Distributions of a mismatched beam (dashed ellipses) at the QF center of a FODO cell. Left: at injection; right: one period later. Solid ellipses are defined by the periodic lattice.

In an accelerator, the ellipse representing the beam distribution is often the same as the ellipse defined by the periodic lattice. In circular accelerators, the beam tends to settle, due to various diffusion processes, to an equilibrium distribution that matches the linear optics functions. For example, in electron storage rings, the equilibrium distribution is determined by radiation damping and quantum diffusion.

When a beam is injected into a machine with a periodic lattice, such as a long linac or a storage ring, the distribution of the incoming beam might not match the ellipse determined by the periodic optics. With such an "optics mismatch", the beam distribution at a periodic observation point will exhibit oscillatory behavior. Figure 1.10 illustrates the beam distribution at two successive passes of the QF center in a FODO cell for a mismatched beam.

1.3 LONGITUDINAL DYNAMICS

In the previous sections we studied the transverse beam motion. The central theme of the transverse beam dynamics is the alternating gradient, strong focusing scheme which provides stability in both x and y planes. In the longitudinal direction, there is also a requirement for stability. The longitudinal stability is provided by radio-frequency (RF) focusing, which is to be discussed in this section.

The principle of RF focusing is based on two aspects: (1) the revolution time of an off-momentum particle differs from that of the reference particle (on-momentum) and the difference is proportional to the momentum deviation; (2) the RF cavities exchange energy with the particle and the energy gain by the particle depends on its arrival time. The reference particle, referred to as the synchronous particle, is chosen to be an imaginary particle that arrives at the RF cavities with the same RF phase on every pass. It must have a specific combination of energy and RF phase to remain synchronous.

The revolution time of a particle is determined by the path length of one revolution and its speed, $T = \frac{C}{v}$, hence

$$\frac{\Delta T}{T} = \frac{\Delta C}{C} - \frac{\Delta v}{v} = (\alpha_c - \frac{1}{\gamma^2})\delta, \qquad (1.84)$$

where γ is the Lorentz energy factor, $\delta = \frac{\Delta p}{p}$, and $\alpha_c = \frac{\Delta C}{\delta C}$ is the momentum compaction factor. The ratio between $\frac{\Delta T}{T}$ and δ is the phase slippage factor,

$$\eta = \alpha_c - \frac{1}{\gamma^2}. \qquad (1.85)$$

When the beam energy satisfies $\gamma = \gamma_T \equiv \frac{1}{\sqrt{\alpha_c}}$, the revolution time does not depend on the momentum deviation to the first order. The corresponding energy factor, γ_T, is referred to as the transition gamma. There is no longitudinal focusing in this case as the phase coordinate is "frozen". If $\eta < 0$, higher energy particles complete a revolution in less time than the synchronous particle, and the opposite is true for $\eta > 0$. The $\eta < 0$ case is said to be below transition, and $\eta > 0$ is above transition.

There is an oscillating electric field along the longitudinal direction across the gap of an RF cavity. This electric field accelerates or decelerates the particles by adding or removing energy from them, respectively, as they traverse the cavity gap. The net effect of the electric field across the gap can be characterized by an oscillating voltage

$$V(t) = V_g \sin(\omega_{\rm rf} t + \phi_s), \qquad (1.86)$$

where $\omega_{\rm rf}$ is the angular resonant frequency of the cavity, ϕ_s is the RF phase when the synchronous particle crosses the gap, and V_g is the RF gap voltage. V_g is related to the peak electric field, $\mathcal{E}$, and the gap length g via

$$V_g = \int_{-g/2}^{g/2} \mathcal{E} \cos\frac{\omega_{\rm rf} z}{\beta c} dz. \qquad (1.87)$$

The resonant frequency of the RF cavity must be a multiple of the revolution frequency, $\omega_{\rm rf} = h\omega_0$, with $\omega_0 = 2\pi/T_0$. T_0 is the revolution time for the reference particle and h is an integer which is defined as the harmonic number.

In a synchrotron the beam energy changes with time as the beam picks up energy from the RF cavity. The revolution frequency may vary with beam energy for low or medium energy proton or heavy ion synchrotrons, which requires the RF frequency to change accordingly. The RF voltage may also ramp with time as the beam energy changes.

The longitudinal motion of an arbitrary particle can be described by its energy and the RF phase at the time it traverses the RF cavity. The total energy change of the particle in one revolution is determined by the energy gain or loss in the RF cavity along with other energy losses, such as due to radiation or coupling impedances. The energy of the synchronous particle on successive turns satisfies

$$E_{s,n+1} = E_{s,n} + eV_n \sin \phi_s - U_{0,n}, \tag{1.88}$$

where subscripts n, $n + 1$ are turn numbers, V is the RF voltage, and U_0 contains all other energy losses. Similarly, the energy of a given particle over the two turns satisfies

$$E_{n+1} = E_n + eV_n \sin \phi_n - U_{0,n}, \tag{1.89}$$

where $\phi_n = \phi_s + \omega_{\mathrm{rf}}\tau_n$, with the arrival time error $\tau = T - T_s$ and T_s the arrival time of the synchronous particle. Therefore, the energy errors, $\Delta E = E - E_s$, of this particle over two successive turns are related through

$$\Delta E_{n+1} = \Delta E_n + eV_n[\sin(\phi_s + \omega_{\mathrm{rf},n}\tau_n) - \sin \phi_s]. \tag{1.90}$$

On the other hand, the arrival time errors of the particle on the two successive turns are related through

$$\tau_{n+1} = \tau_n + \Delta T_{n+1} = \tau_n + \eta \delta T_{0,n+1},$$

where ΔT_{n+1} is the arrival time error accumulated over turn $n + 1$. Using $\omega_{\mathrm{rf}}T_0 = 2\pi h$ and $\delta = \frac{1}{\beta^2}\frac{\Delta E}{E}$, the above equation can be rewritten in terms of the energy error coordinate,

$$\tau_{n+1} = \tau_n + \frac{2\pi h \eta(E_{s,n+1})}{\omega_{\mathrm{rf},n+1}\beta_{s,n+1}^2 E_{s,n+1}}\Delta E_{n+1}. \tag{1.91}$$

Eqs. (1.90) and (1.91) are the two mapping equations that describe the longitudinal motion of particles in a synchrotron in $(\tau, \Delta E)$ coordinates. Given the ramping curves of the reference beam energy, the RF voltage, and the RF frequency, the trajectory of a particle in the $(\tau, \Delta E)$ phase space can be determined by applying the equations over successive turns. Eq. (1.90) describes the energy change at the RF cavity, and Eq. (1.91) describes the arrival time change accumulated throughout the ring. They can be seen as two maps that are applied sequentially. In each revolution one first calculates the energy change at the cavity, followed by the calculation of the arrival time change in the ring. Both maps are kick maps, i.e., maps in which an increment is made only to one of the two conjugate coordinates. The determinant of the Jacobian matrix of a kick map is unity, which means that the phase space area of a beam distribution is preserved as it evolves with time.

The stability of the particle motion around the reference particle can be analyzed after linearizing Eq. (1.90) with respect to the τ coordinate, which becomes

$$\Delta E_{n+1} = \Delta E_n + \tau_n \omega_{\mathrm{rf},n} eV_n \cos \phi_s. \tag{1.92}$$

Combining Eqs. (1.91)-(1.92), the linear longitudinal motion can be described by a one-turn transfer matrix,

$$
\begin{pmatrix} \tau_{n+1} \\ \Delta E_{n+1} \end{pmatrix} = \begin{pmatrix} 1 + \frac{2\pi h \eta e V \cos \phi_s}{\beta^2 E} & \frac{2\pi h \eta}{\omega_{\mathrm{rf}} \beta^2 E} \\ \omega_{\mathrm{rf}} e V \cos \phi_s & 1 \end{pmatrix} \begin{pmatrix} \tau_n \\ \Delta E_n \end{pmatrix}, \tag{1.93}
$$

where we have dropped the subscripts for parameters in the transfer matrix for notation simplicity. The stability of motion requires the trace of the transfer matrix to be between -2 and 2 (see Eq. (1.55)), hence

$$
-4 < \frac{2\pi h \eta e V \cos \phi_s}{\beta^2 E} < 0. \tag{1.94}
$$

In reality, the value of the RF voltage is low compared to the beam energy in electron-volts, such that $|2\pi h \eta e V / E| \ll 1$. Therefore, to ensure the longitudinal stability, one only needs to choose the RF phase according to the sign of the phase slippage factor such that

$$
\eta \cos \phi_s < 0, \tag{1.95}
$$

is satisfied.

Within the stability region, the longitudinal coordinates oscillate about the synchronous particle. The stable oscillation in the longitudinal direction is called synchrotron motion. The synchrotron tune, defined as the number of oscillations per turn, can be found with the trace of the transfer matrix in the same manner as the analysis of transverse motion.

Another commonly used set of coordinates for the longitudinal motion is the (ϕ, δ) coordinates, in which the mapping equations become

$$
\delta_{n+1} = \frac{\beta_{s,n}^2 E_{s,n}}{\beta_{s,n+1}^2 E_{s,n+1}} \delta_n + \frac{e V_n}{\beta_{s,n+1}^2 E_{s,n+1}} (\sin \phi_n - \sin \phi_s), \tag{1.96a}
$$

$$
\phi_{n+1} = \frac{\omega_{\mathrm{rf},n+1}}{\omega_{\mathrm{rf},n}} \phi_n + 2\pi h \eta (\delta_{n+1}) \delta_{n+1}. \tag{1.96b}
$$

In this case the determinant of the Jacobian matrix for the transformation from turn n to $n+1$ is

$$
\left\| \frac{\partial(\delta_{n+1}, \phi_{n+1})}{\partial(\delta_n, \phi_n)} \right\| = \frac{\beta_{s,n}^2 E_{s,n}}{\beta_{s,n+1}^2 E_{s,n+1}} \frac{\omega_{\mathrm{rf},n+1}}{\omega_{\mathrm{rf},n}}. \tag{1.97}
$$

The non-unity result means that the phase space area of a closed contour in (ϕ, δ) coordinates will change. Using Eq. (1.97), it is easy to show that the phase space area in (ϕ, δ) coordinates, A, scales with the beam energy and the RF frequency such that

$$
A(\phi, \delta) \frac{\beta^2 E}{\omega_{\mathrm{rf}}} = \mathrm{const.} \tag{1.98}
$$

In a storage ring, beam energy and the RF frequency do not change. The RF voltage is typically also held constant. The mapping equations are simplified to

$$\delta_{n+1} = \delta_n + \frac{eV}{\beta^2 E}(\sin\phi_n - \sin\phi_s), \tag{1.99a}$$

$$\phi_{n+1} = \phi_n + 2\pi h\eta(\delta_{n+1})\delta_{n+1}. \tag{1.99b}$$

Usually the synchrotron motion is slow (one synchrotron oscillation takes tens to hundreds of turns), hence the discrete motion described by the above difference equations can be approximated with a smooth, continuous motion described by differential equations. Using the average values in one turn to approximate the derivatives,

$$\dot{\delta} \equiv \frac{d\delta}{dt} \approx \frac{\delta_{n+1} - \delta_n}{T_0}, \qquad \dot{\phi} \equiv \frac{d\phi}{dt} \approx \frac{\phi_{n+1} - \phi_n}{T_0}, \tag{1.100}$$

Eqs. (1.99a-1.99b) become

$$\dot{\delta} = \frac{\omega_0 eV}{2\pi\beta_s^2 E_s}(\sin\phi - \sin\phi_s), \qquad \dot{\phi} = h\omega_0\eta\delta. \tag{1.101}$$

These equations can be derived from the Hamiltonian

$$H = \frac{1}{2}h\omega_0\eta\delta^2 + \frac{\omega_0 eV}{2\pi\beta_s^2 E_s}\left[\cos\phi - \cos\phi_s + (\phi - \phi_s)\sin\phi_s\right], \tag{1.102}$$

with canonical coordinates (ϕ, δ) and free variable t. Eq. (1.102) is the Hamiltonian for synchrotron motion.

The characteristics of the synchrotron motion can be studied with the Hamiltonian. The motion of a particle with any given initial condition will follow a path in the phase space determined by the Hamiltonian. For the Hamiltonian in Eq. (1.102), there are two fixed points in the phase space, which can be determined from Eq. (1.101) by setting $\dot{\delta} = \dot{\phi} = 0$. Under the condition $\eta\cos\phi_s < 0$, the fixed point at $(\phi = \phi_s, \delta = 0)$ is stable, and the other fixed point, located at $(\phi = \pi - \phi_s, \delta = 0)$, is unstable. When a particle is launched with an initial condition in the vicinity of the stable fixed point, it will move on an ellipse centered at the fixed point. On the other hand, the Hamiltonian contour that passes the unstable fixed point defines the boundary of stable and unstable motion. This contour, defined by the equation,

$$H(\phi, \delta) = H(\pi - \phi_s, 0), \tag{1.103}$$

is referred to as the separatrix. The equation can be written as

$$\text{sgn}(\eta)\bar{\delta}^2 + \cos\phi + \phi\sin\phi_s = -\cos\phi_s + (\pi - \phi_s)\sin\phi_s, \tag{1.104}$$

with the normalized momentum deviation coordinate defined by

$$\bar{\delta} = \delta\sqrt{\frac{\pi h|\eta|\beta_s^2 E_s}{eV}}. \tag{1.105}$$

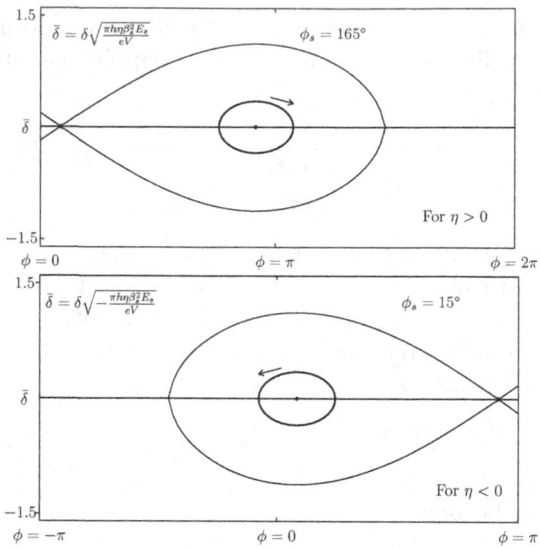

Figure 1.11 Examples of RF buckets above and below transition. Top: ($\eta > 0$, $\phi_s = 165°$), above transition; Bottom: ($\eta < 0$, $\phi_s = 15°$), below transition. The small ellipses represent the stable synchrotron motion.

The closed area within the separatrix is the stable region where beam can be stored or accelerated. This area is called the RF bucket. Outside of the RF bucket, particles will fall out of phase with the RF waveform and eventually get lost. The RF buckets for the two cases, below and above transition, are shown in Figure 1.11.

The height, length, and area of the RF buckets are important measures of the stability region. These can all be derived from Eq. (1.104). The bucket length is the distance from the unstable fixed point to the other end of the bucket, the point (ϕ_u, 0), where ϕ_u can be found with

$$\cos \phi_u + \phi_u \sin \phi_s = -\cos \phi_s + (\pi - \phi_s) \sin \phi_s. \tag{1.106}$$

The bucket height is the maximum momentum deviation on the separatrix, which is given by

$$\bar{\delta}_m = \sqrt{2} \left| \cos \phi_s + (\phi_s - \frac{\pi}{2}) \sin \phi_s \right|^{\frac{1}{2}}. \tag{1.107}$$

The bucket area is given by [78]

$$\bar{\mathcal{A}} = \oint_{spx} \bar{\delta} d\phi = 8\sqrt{2}\alpha(\phi_s), \quad \text{with } \alpha(\phi_s) \approx \frac{1 - \sin \phi_s}{1 + \sin \phi_s}. \tag{1.108}$$

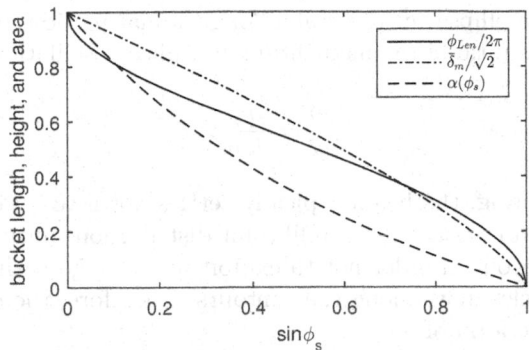

Figure 1.12 The RF bucket length, height, and area as represented by scaled parameters $\phi_{\text{Len}}/2\pi$, $\bar{\delta}_m/\sqrt{2}$, and $\alpha(\phi_s)$, respectively, with $\phi_{\text{Len}} = |\phi_s + \phi_u - \pi|$.

The dimensions of the RF buckets as a function of $\sin\phi_s$ are given in Figure 1.12. The bucket size is at its maximum when the synchronous phase is $\phi_s = 0$ or π. In this case, there is no net energy transfer to the beam; such an RF bucket is called the stationary bucket.

With sufficiently small oscillation amplitudes, the synchrotron motion around the stable fixed point is essentially linear. Introducing the phase deviation variable

$$\varphi = \phi - \phi_s, \tag{1.109}$$

the Hamiltonian can be expanded to give

$$H = \frac{1}{2} h\omega_0 \eta \delta^2 - \frac{\omega_0 eV \cos\phi_s}{4\pi\beta_s^2 E_s} \varphi^2, \tag{1.110}$$

when keeping only the lowest order terms. The equation of motion of this Hamiltonian is

$$\ddot{\varphi} + \omega_s^2 \varphi = 0, \tag{1.111}$$

with $\omega_s = \nu_s\omega_0$, and ν_s is the synchrotron tune, given by

$$\nu_s = \left(\frac{heV|\eta\cos\phi_s|}{2\pi\beta_s^2 E_s} \right)^{1/2}. \tag{1.112}$$

The solution to Eq. (1.111) is

$$\varphi = \hat{\varphi}\cos(\omega_s t + \chi), \qquad \delta = -\text{sgn}(\eta)\hat{\delta}\sin(\omega_s t + \chi), \tag{1.113}$$

where $\hat{\varphi}$ and $\hat{\delta}$ are the oscillation amplitudes in the φ, δ directions, respectively. The phase space ellipses of the stable longitudinal motion are illustrated in Figure 1.11. The ratio of the momentum and phase oscillation amplitudes is

$$\frac{\hat{\delta}}{\hat{\varphi}} = \frac{\nu_s}{h|\eta|}. \tag{1.114}$$

In a storage ring, the beam typically settles down to an equilibrium distribution. In phase space, the equilibrium distribution must conform to the Hamiltonian contour in order not to exhibit any change with time while the individual particles move along the contours. Therefore, the equilibrium distribution has the form of

$$\rho(\varphi, \delta) \sim \rho(H(\varphi, \delta)) \sim \rho(\hat{\varphi}), \tag{1.115}$$

where $\hat{\varphi}$ is related to the Hamiltonian through $H = \frac{1}{2} \frac{\omega_0 \nu_s^2}{h\eta} \hat{\varphi}^2$. In an electron storage ring, the combination of radiation induced damping and diffusion creates a Gaussian distribution, which can be written as

$$\rho(\hat{\varphi}) = \frac{1}{2\pi\sigma_{\hat{\varphi}}^2} \exp\left(-\frac{1}{2}\frac{\hat{\varphi}^2}{\sigma_{\hat{\varphi}}^2}\right), \tag{1.116}$$

or expressed in the phase space coordinates ϕ and δ, rather than the phase amplitude $\hat{\phi}$,

$$\rho(\varphi, \delta) = \frac{1}{2\pi\sigma_\varphi\sigma_\delta} \exp\left[-\frac{1}{2}\left(\frac{\varphi^2}{\sigma_\varphi^2} + \frac{\delta^2}{\sigma_\delta^2}\right)\right], \tag{1.117}$$

where σ_φ is the rms bunch length and σ_δ is the rms momentum deviation.

Beam dynamics topics

CONTENTS

In Chapter 1 we studied the linear beam motion in an ideal lattice. A real accelerator has all sorts of errors, which cause the beam motion to deviate from the design, in terms of the beam orbit, linear optics, and linear coupling. The intrinsic field errors in dipole and quadrupole magnets for off-energy particles introduce dispersion and chromatic aberration. The use of sextupole magnets to correct chromaticity in storage rings makes the beam motion nonlinear. These effects are discussed in this chapter. The accurate modeling of accelerators with simulation codes is also discussed.

2.1   BEAM ORBIT

In the ideal scenario, the magnetic fields along the beam path are identical to what is specified in the design, and hence a beam launched on the design orbit will travel on the design orbit. However, the actual field distribution on the beam path will always be different from the design. The differences are the magnetic field errors, which can be due to magnet imperfections, calibration errors, power supply regulation errors, power grid ripples, and magnet misalignment. A major impact of the field errors to the beam is to cause the actual beam orbit to deviate from the design. Correction of orbit errors is a basic requirement in accelerator operation.

Magnet misalignment introduces unintended multipole field components on the beam through the so-called "feed-down" effects. Quadrupole and sextupole magnets are typically placed to have the magnetic axes on the design orbit, such that a particle traveling on the design orbit sees no magnetic field. When the magnet centers are shifted transversely to $(-x_0, -y_0)$, so that the point (x, y) has coordinates relative to the magnet center

$$X = x_0 + x, \qquad Y = y_0 + y,$$

the magnetic fields in a quadrupole at the point will be

$$B_y = b_1 X = b_1 x_0 + b_1 x, \quad B_x = b_1 Y = b_1 y_0 + b_1 y, \tag{2.1}$$

and similarly the field in a sextupole will be

$$B_y = \frac{b_2}{2} \left((x_0^2 - y_0^2) + (2x_0 x - 2y_0 y) + (x^2 - y^2) \right), \tag{2.2}$$

$$B_x = b_2 (x_0 y_0 + x y_0 + x_0 y + xy). \tag{2.3}$$

The constant terms (independent of x and y) in the B_y and B_x are dipole field errors. Dipole field errors give the beam angular kicks (i.e., changes to x' and y' coordinates),

$$\theta_x = \frac{B_y \Delta s}{B\rho}, \quad \theta_y = -\frac{B_x \Delta s}{B\rho}, \tag{2.4}$$

where Δs is length of the field error. The angular kicks propagate downstream and alter the beam orbit.

In a one-pass system (linac or transport line), the beam orbit change due to a kick is calculated with the transfer matrix between the kick and the observation points. Initial orbit errors at the entrance point of the line will also propagate through the transfer matrix. The overall orbit error at any location, P, can be obtained by summing up contributions from all upstream kicks and the effects of the initial steering errors

$$\Delta\mathbf{X} = \mathbf{M}(P|0)\mathbf{X}_0 + \sum_i \mathbf{M}(P|i)\boldsymbol{\theta}_i, \quad \text{with} \quad \boldsymbol{\theta}_i = \begin{pmatrix} 0 \\ \theta_{xi} \\ 0 \\ \theta_{yi} \end{pmatrix}, \tag{2.5}$$

where $\mathbf{M}(P|0)$ is the transfer matrix from the entrance to point P, $\mathbf{M}(P|i)$ is the transfer matrix from location i to P, $\mathbf{X}_0$ is the launching orbit, and θ_{xi} and θ_{yi} are the horizontal and vertical angular kicks at location i, respectively. Figure 2.1 shows the trajectory deviation due to a vertical kick in the Linac Coherent Light Source (LCLS) as an example.

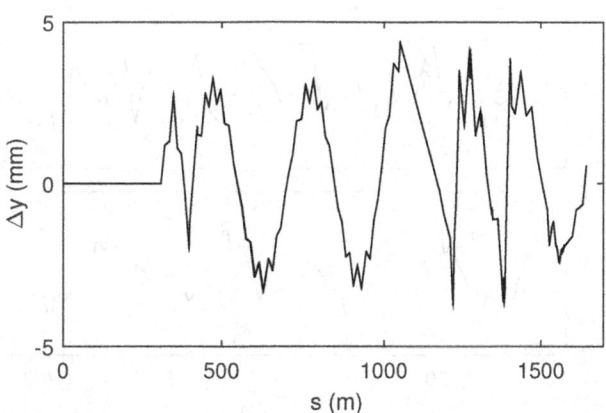

Figure 2.1 Trajectory deviation in a section of LCLS due to a vertical kick of 0.1 mrad.

In a circular accelerator, the orbit kicks from persistent dipole field errors act on the beam on every revolution. The net effect is that the stationary orbit, or closed-orbit, is changed. The closed orbit is a fixed point of the one-turn transfer map, $\mathbf{M}$, which satisfies

$$\mathbf{X}_{co} = \mathbf{M}(\mathbf{X}_{co}). \tag{2.6}$$

Up to the linear order, the closed-orbit condition is (see Eq. (1.48))

$$\mathbf{X}_{co} = \mathbf{M}\mathbf{X}_{co} + \mathbf{\Delta}, \tag{2.7}$$

from which the closed orbit can be solved, giving

$$\mathbf{X}_{co} = (\mathbf{I} - \mathbf{M})^{-1}\mathbf{\Delta}. \tag{2.8}$$

Eq. (2.8) can be used to calculate the orbit response due to a local angular kick. Using the Courant-Snyder parametrization, Eq. (1.58), for the 2×2 one-turn transfer matrix at the point immediately downstream of the kick θ, the closed orbit is found to be

$$\mathbf{X}_{co} = \begin{pmatrix} 1 - \cos\Phi - \alpha_0 \sin\Phi & -\beta_0 \sin\Phi \\ \gamma_0 \sin\Phi & 1 - \cos\Phi + \alpha_0 \sin\Phi \end{pmatrix}^{-1} \begin{pmatrix} 0 \\ \theta \end{pmatrix},$$

where $\Phi = 2\pi\nu$, ν is the betatron tune, and thus

$$\mathbf{X}_{co} = \begin{pmatrix} y_0 \\ y_0' \end{pmatrix} = \frac{\theta}{2 \sin\pi\nu} \begin{pmatrix} \beta_0 \cos\pi\nu \\ \sin\pi\nu - \alpha_0 \cos\pi\nu \end{pmatrix}. \tag{2.9}$$

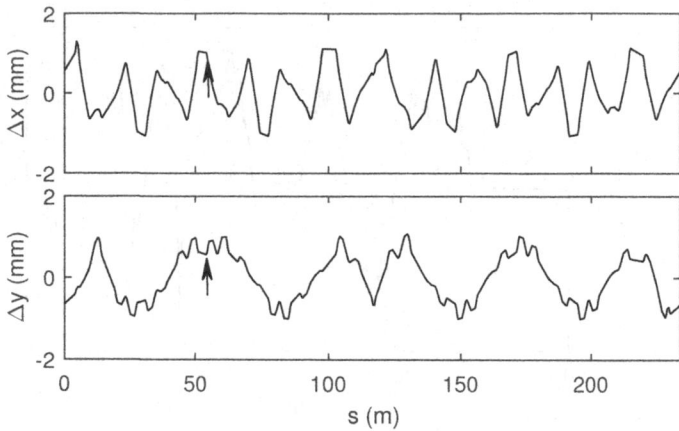

Figure 2.2 Horizontal (top) and vertical (bottom) closed orbit deviation in SPEAR3 ($\nu_x = 14.106$, $\nu_y = 6.177$) by one kick of 0.1 mrad in each plane. The location of the kick is marked by the arrows.

The closed orbit at other locations can be found with the transfer matrix. The closed orbit position coordinate at location s is given by

$$y_{co}(s) = \frac{\theta\sqrt{\beta(s)\beta_0}}{2\sin\pi\nu}\cos(|\psi(s) - \psi_0| - \pi\nu), \qquad (2.10)$$

where β_0 and ψ_0 are beta function and betatron phase advance at the location of the kick, respectively. Starting from the location of the kick, the orbit varies as a sinusoidal function of the betatron phase advance with its amplitude scaled by the factor $\sqrt{\beta(s)}$.

The closed orbit shift due to a localized angular kick by an orbit corrector is called the orbit response of the corrector. As shown in Eq. (2.10), the orbit response is closely related to the linear optics. Therefore, the measured orbit responses can be used to determine the linear optics errors of the ring. These are to be discussed in Chapter 4. Figure 2.2 shows the closed orbit deviation in the SPEAR3 storage ring due to an orbit kick in each of the transverse planes.

In reality there would be many sources of dipole field errors distributed around the ring that give kicks to the beam. Considering only the linear lattice, the closed orbit is the sum of the contribution of all errors, which can be written as [26]

$$y_{co}(s) = \frac{\sqrt{\beta(s)}}{2\sin\pi\nu}\int_s^{s+C}\theta(s')\sqrt{\beta(s')}\cos(\pi\nu + \psi(s) - \psi(s'))ds', \qquad (2.11)$$

where $\theta(s')ds$ represents the kick from the section between s and $s + ds$.

Introducing the transformation $\psi(s) = \nu\phi(s)$, and using $ds = \beta d\psi = \beta\nu d\phi$, the closed orbit, Eq. (2.11) can be rewritten as

$$y_{co}(s) = \frac{\nu\sqrt{\beta(s)}}{2\sin\pi\nu} \int_\phi^{\phi+2\pi} \theta(\xi)\beta^{\frac{3}{2}}(\xi)\cos\nu(\pi + \phi - \xi)d\xi. \tag{2.12}$$

Because the function $\theta(\phi)\beta^{\frac{3}{2}}(\phi)$ is periodic (with the period 2π), it can be Fourier expanded

$$\theta(\phi)\beta^{\frac{3}{2}}(\phi) = \sum_{n=-\infty}^{\infty} f_n e^{in\phi}, \tag{2.13}$$

with the Fourier coefficients f_n given by

$$f_n = \frac{1}{2\pi\nu} \oint \theta(s)\beta^{\frac{1}{2}}(s)e^{-in\phi(s)}ds = \frac{1}{2\pi\nu} \sum_k \theta_k \beta_k^{\frac{1}{2}} e^{-in\phi_k}, \tag{2.14}$$

where in the last equation discrete dipole kicks are assumed, with a kick angle $\theta_k = \theta(s_k)\Delta s_k$. The Fourier coefficients in Eq. (2.14) are called integer stopband integrals. Inserting Eq. (2.13) into Eq.(2.12), we obtain the Fourier expansion of the closed orbit deviation

$$y_{co}(s) = \sqrt{\beta(s)} \sum_n \frac{\nu^2 f_n}{\nu^2 - n^2} e^{in\phi(s)},$$

$$= \sqrt{\beta(s)} \left(f_0 + \sum_{n=1}^{\infty} \frac{2\nu^2|f_n|}{\nu^2 - n^2} \cos(n\phi(s) + \chi_n) \right), \tag{2.15}$$

where we have used $f_n = f_{-n}^* = |f_n|e^{i\chi_n}$. The closed orbit can be approximated with only a few Fourier terms with n near the betatron tune, ν. Keeping the leading term only, we have an approximation

$$y_{co}(s) \approx \sqrt{\beta(s)} \frac{\nu|f_{[\nu]}|}{\nu - [\nu]} \cos([\nu]\phi(s) + \chi_{[\nu]}), \tag{2.16}$$

where $[\nu]$ is the integer closest to ν.

Eq. (2.16) shows that, unless the orbit kicks are specifically arranged, the closed orbit deviation in a ring accelerator tends to appear as an amplitude modulated sinusoidal function of the phase advance, with the number of peaks around the circumference given by the integer part of the betatron tune. It also indicates that if the betatron tune is close to an integer, a small perturbation could cause very large orbit deviations.

Orbit errors of the beam in accelerators generally need to be corrected. Depending on the purpose of the accelerator, the precision requirement for orbit correction may vary. The orbit at the interaction points of colliders or at the photon beamline source points of light sources needs to be held constant with a high precision at the sub-micron level for hours.

2.2 LINEAR OPTICS ERRORS

In this section we discuss the effects of quadrupole field errors [26]. Since quadrupole magnets determine the linear optics of the accelerators, deviations of the quadrupole fields from the design will distort the linear optics. In other words, the propagation of phase space coordinates through the beam line will be changed.

We consider a localized quadrupole field error, which can be represented by a thin-lens quadrupole. Its transfer matrix can be written as

$$\mathbf{M_q} = \begin{pmatrix} 1 & 0 \\ -k_0 & 1 \end{pmatrix}, \tag{2.17}$$

where $k_0 = k(s)\Delta s$ is the integrated gradient error, with $k(s)$ the gradient error and Δs the length of the field error.

In a one-pass system, the localized quadrupole error has no impact to the beam motion upstream. It also has no impact to the propagation of orbit errors located downstream. However, it will affect how coordinate deviations upstream of the error are propagated to downstream locations. This effect is described by the change to the transfer matrix. Suppose point 1 is upstream of the quadrupole error and point 2 is downstream. The transfer matrix from point 1 to 2 without the quadrupole error is $\mathbf{M_0}(s_2|s_1) = \mathbf{M}(s_2|s_q)\mathbf{M}(s_q|s_1)$. With the quadrupole field error, it becomes $\mathbf{M}(s_2|s_1) = \mathbf{M}(s_2|s_q)\mathbf{M_q}\mathbf{M}(s_q|s_1)$. The changes to the transfer matrix due to the quadrupole error are

$$\mathbf{M}(s_2|s_1) - \mathbf{M_0}(s_2|s_1) = -k_0 \begin{pmatrix} M_{12}^{(2)} M_{11}^{(1)} & M_{12}^{(2)} M_{12}^{(1)} \\ M_{22}^{(2)} M_{11}^{(1)} & M_{22}^{(2)} M_{12}^{(1)} \end{pmatrix}, \tag{2.18}$$

where $\mathbf{M}^{(1)} = \mathbf{M}(s_q|s_1)$ and $\mathbf{M}^{(2)} = \mathbf{M}(s_2|s_q)$. With the changes to the transfer matrix, if the optics functions are specified at the point 1, the changes to the optics functions at point 2 can be calculated using Eq. (1.70). A quadrupole error in a one-pass system only affects the optics functions downstream.

In a circular accelerator, since the optics functions are derived from the parametrization of the one-turn transfer matrix and a quadrupole error anywhere in the ring will change the one-turn transfer matrix, a quadrupole error changes the optics functions everywhere. For example, right at the downstream face of the quadrupole error, the one-turn transfer matrix becomes

$$\mathbf{M} = \mathbf{M_q}\mathbf{M_0} = \begin{pmatrix} 1 & 0 \\ -k_0 & 1 \end{pmatrix} \begin{pmatrix} \cos\Phi_0 + \alpha_0 \sin\Phi_0 & \beta_0 \sin\Phi_0 \\ -\gamma_0 \sin\Phi_0 & \cos\Phi_0 - \alpha_0 \sin\Phi_0 \end{pmatrix}. \tag{2.19}$$

From the trace of matrix $\mathbf{M}$, we find

$$\cos\Phi - \cos\Phi_0 = -\frac{1}{2}k_0\beta_0 \sin\Phi_0, \tag{2.20}$$

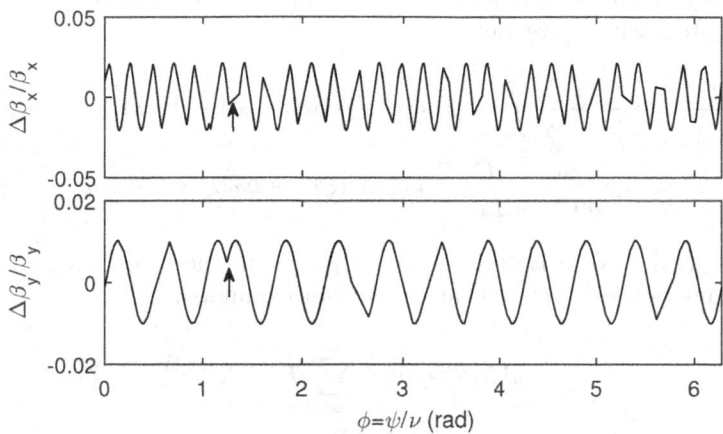

Figure 2.3 Beta beating for SPEAR3 by one quadrupole error (+0.5% gradient error on a focusing quadrupole, location marked by arrows).

and hence the change of betatron tune is given by

$$\Delta\nu \equiv \nu - \nu_0 = \frac{1}{2\pi}(\Phi - \Phi_0) \approx \frac{1}{4\pi}k_0\beta_0. \tag{2.21}$$

For quadrupole errors distributed around the ring, with gradient errors given by $k(s)$, the total tune change can be obtained by integrating the contributions of all errors, which leads to the formula

$$\Delta\nu = \frac{1}{4\pi}\oint k(s)\beta(s)ds. \tag{2.22}$$

Parametrization of matrix $\mathbf{M}$ yields the new Courant-Snyder parameters at the exit face of the quadrupole error. Courant-Snyder parameters at other locations can be obtained by the propagation formula, Eq. (1.70). The fractional change to the beta function is found to be

$$\frac{\Delta\beta(s)}{\beta(s)} = -\frac{k_0\beta_0}{2\sin\Phi_0}\cos(2|\psi(s) - \psi(s_0)| - \Phi_0). \tag{2.23}$$

The fractional deviation of beta function from the design, $\frac{\Delta\beta}{\beta}$, is referred to as beta beating. Eq. (2.23) shows that the beta beating caused by a single quadrupole error is a sinusoidal function that propagates at twice the frequency of betatron oscillation. Figure 2.3 shows the beta beating for SPEAR3 due to a localized quadrupole error.

For distributed quadrupole errors, the contributions from all error sources are integrated, which give [26]

$$
\begin{aligned}
\frac{\Delta\beta(s)}{\beta(s)} &= -\frac{1}{2\sin\Phi_0} \int_s^{s+C} k(s')\beta(s') \cos(2\psi(s') - 2\psi(s) - \Phi_0) ds', \\
&= -\frac{\nu_0}{2\sin 2\pi\nu_0} \int_\phi^{\phi+2\pi} k(\xi)\beta^2(\xi) \cos 2\nu_0(\xi - \phi - \pi) d\xi,
\end{aligned}
\tag{2.24}
$$

where we used $\psi(s) = \nu_0\phi(s)$ and $ds = \beta\nu_0 d\phi$ in the second equality. The periodic function $\nu_0 k(\phi)\beta^2(\phi)$ can be Fourier expanded,

$$
\nu_0 k(\phi)\beta^2(\phi) = \sum_{n=\infty}^{\infty} J_n e^{in\phi},
\tag{2.25}
$$

with the Fourier coefficients given by

$$
\begin{aligned}
J_n \equiv |J_n|e^{i\chi} &= \frac{1}{2\pi} \oint \nu_0 k(\phi)\beta^2(\phi)e^{-in\phi} d\phi, \\
&= \frac{1}{2\pi} \oint k(s')\beta(s')e^{-in\phi(s')} ds'.
\end{aligned}
\tag{2.26}
$$

The Fourier coefficients J_n are called half-integer stopband integrals. Inserting Eq. (2.25) into the integral in Eq.(2.24), beta beating can now be expressed in Fourier series as

$$
\frac{\Delta\beta(s)}{\beta(s)} = -\frac{\nu_0}{2} \sum_n \frac{J_n e^{in\phi(s)}}{\nu_0^2 - \frac{1}{4}n^2} = -\frac{J_0}{2\nu_0} + \sum_{n=1}^{\infty} \frac{\nu_0|J_n|}{\frac{1}{4}n^2 - \nu_0^2} \cos(n\phi(s) + \chi_n).
\tag{2.27}
$$

The beta beating in a ring is usually dominated by the Fourier harmonics close to $2\nu_0$. Keeping only the leading term, with $n = [2\nu_0]$, the integer closest to $2\nu_0$, the beta beating is approximately

$$
\frac{\Delta\beta(s)}{\beta(s)} \approx \frac{|J_{[2\nu]}|}{2\nu_0 - [2\nu]} \cos\left([2\nu]\phi(s) + \chi_{[2\nu]}\right).
\tag{2.28}
$$

Eq. (2.28) shows that the betatron tune cannot be too close to a half integer, otherwise the beta beating will diverge. It can be shown that the beam motion in the periodic lattice becomes unstable if the distance between the betatron tune and a half-integer is less than a half of $|J_{[2\nu]}|$, i.e., when

$$
|\nu_0 - \frac{1}{2}[2\nu]| \leq \frac{1}{2}|J_{[2\nu]}|.
\tag{2.29}
$$

$|J_{[2\nu]}|$ is called the width of the half-integer stopband.

The betatron phase advance is closely related to the beta function via $d\psi = \frac{ds}{\beta}$. The errors to the phase advance due to quadrupole errors can be calculated from the beta beating, which gives

$$\Delta\psi = \Delta \int \frac{ds}{\beta} = -\int \frac{\Delta\beta}{\beta^2} ds = -\int \frac{\Delta\beta}{\beta} \nu_0 d\phi,$$

$$= \frac{J_0 \psi(s)}{2\nu_0} + \sum_{n=1}^{\infty} \frac{1}{n} \frac{\nu_0^2 |J_n|}{\nu_0^2 - \frac{1}{4}n^2} \sin(n\phi(s) + \chi_n). \qquad (2.30)$$

The $n = 0$ term gives the betatron tune shift, $\Delta\nu = \frac{J_0}{2}$.

Linear optics errors have many detrimental effects and correction of the optics errors is usually desired. By increasing the beta function at locations of limiting apertures, beta beating may reduce the acceptance of the machine and in turn cause a reduced tuning range, scrape off the beam on the vacuum chamber, or reduce the injection efficiency. Deviations of beta functions from the design may cause radiation protection issues as the beam loss distribution in the machine is changed. In transport lines, errors of optics functions at the extraction point cause a mismatch of the beam distribution with the optics of the downstream accelerator and may reduce the injection efficiency. In a free electron laser (FEL), optics mismatch in the undulator will reduce the FEL power.

In a storage ring, optics errors may significantly impact the nonlinear beam dynamics performance, resulting in a reduced dynamic aperture and local momentum apertures, which may in turn decrease the injection efficiency and the Touschek lifetime, respectively. Typically the nonlinear beam dynamics of a storage ring design makes use of a cancellation scheme of the nonlinear effects by the sextupoles. The cancellation scheme relies on the phase advances between certain pairs of sextupoles being close to an odd multiple of π. As optics errors distort the betatron phase advances, the cancellation scheme does not work as expected. This may lead to increases of the nonlinear resonance strengths and changes to the tune footprint. The excitation of certain nonlinear resonances may reduce the dynamic aperture or local momentum apertures.

2.3 DISPERSION

Dipole fields, including the fields in dipole magnets and the feed-down dipole components in quadrupoles and sextupoles, determine the beam orbit in an accelerator. Because the bending angle of a dipole magnet for a beam is inversely proportional to the beam energy, the beam orbit depends on the beam energy. The dependence of beam orbit on beam energy is called dispersion.

Eqs. (1.29) describes the particle motion through a dipole magnet, including the effect of energy errors. This form of equation can be extended to describe the particle motion in a general accelerator section,

$$\mathbf{X} = \mathbf{M}\mathbf{X}_0 + \delta \mathbf{d}, \qquad (2.31)$$

with phase space coordinate $\mathbf{X} = (x, x')^T$, transfer matrix $\mathbf{M}$, and $\mathbf{d}$ a 2-element column vector such that $\delta\mathbf{d}$ represents the accumulated orbit shift induced by the momentum deviation over the section. The vertical plane is not considered here because there is no vertical dispersion in an ideal planar accelerator. The same treatment can be extended to include the vertical plane.

Eq. (2.31) shows that a part of the phase space coordinate, $\mathbf{X}$, is linearly dependent on the momentum deviation. It is possible and desirable to separate this part out from the usual betatron motion, with

$$\mathbf{X} = \mathbf{X}_\beta + \mathbf{D}\delta, \tag{2.32}$$

where the dispersion vector $\mathbf{D}=(D, D')^T$, and D is the dispersion function, $D' = \frac{dD}{ds}$. Inserting Eq. (2.32) into Eq. (2.31) and separating the terms dependent on δ, we obtain

$$\mathbf{X}_\beta = \mathbf{M}\mathbf{X}_{\beta,0}, \tag{2.33}$$

$$\mathbf{D} = \mathbf{M}\mathbf{D}_0 + \mathbf{d}, \tag{2.34}$$

where $\mathbf{X}_0 = \mathbf{X}_{\beta,0} + \mathbf{D}_0\delta$ was used. The $\mathbf{X}_\beta$ term represents the betatron motion and is called the betatron coordinate.

Eq. (2.34) specifies how the dispersion vector is transported through an accelerator section. At the exit face the dispersion vector consists of a term that is transported from the initial dispersion vector by the transfer matrix and a term that represents the contributions from the dipole fields in the section itself.

In a one-pass system, the choice of the initial values of the dispersion functions is somewhat arbitrary. Usually the initial dispersion is chosen to match the expected beam distribution at the entrance point. In this case, the dispersion vector obtained with Eq. (2.34) will be consistent with the dispersion derived from the beam distribution at a downstream location. For a transport line, the initial dispersion is usually given by the dispersion function values at the extraction point of the upstream machine. For a linac, the initial dispersion may be set to zero, assuming the initial transverse distribution has no correlation with the momentum deviation. In general, given an initial distribution, $f(\mathbf{X}, \delta)$, the dispersion function can be found from

$$D\sigma_\delta^2 = \int \delta x f(\mathbf{X}, \delta) d\mathbf{X} d\delta, \quad D'\sigma_\delta^2 = \int \delta x' f(\mathbf{X}, \delta) d\mathbf{X} d\delta, \tag{2.35}$$

where $\sigma_\delta^2 = \int \delta^2 f(\mathbf{X}, \delta) d\mathbf{X} d\delta$ and σ_δ is defined as the rms momentum spread.

In a circular accelerator, there is a natural choice for the separation of the betatron coordinates and the dispersion terms, which is to impose the periodic condition on the dispersion function,

$$D(s + C) = D(s), \quad D'(s + C) = D'(s), \tag{2.36}$$

where C is the circumference. Applying the periodic condition to Eq. (2.34), with $\mathbf{M}$ being the one-turn transfer matrix and $\mathbf{d}$ the derivative of the orbit shift with respect to δ, the dispersion vector is found to be

$$\mathbf{D} = (\mathbf{I} - \mathbf{M})^{-1}\mathbf{d}. \tag{2.37}$$

The periodic dispersion function in circular accelerators is essentially the derivative of the closed orbit for an off-momentum particle with respect to the momentum deviation, i.e.,

$$\mathbf{D} = \frac{d\mathbf{X}_c(\delta)}{d\delta}, \tag{2.38}$$

where the closed orbit satisfies the periodic condition

$$\mathbf{X}_c(\delta) = \mathbf{M}\mathbf{X}_c(\delta) + \delta\mathbf{d}. \tag{2.39}$$

The calculation of $\mathbf{d}$ can be done by sequentially applying Eq. (2.31) to all elements of the ring. This calculation can be facilitated by using the extended transfer matrix for the $\mathbf{X}_e = (x, x', \delta)^T$ coordinates, for which Eq. (2.31) can be rewritten as

$$\mathbf{X}_e = \mathbf{M}_e\mathbf{X}_{e,0}, \quad \text{with } \mathbf{M}_e = \begin{pmatrix} \mathbf{M} & \mathbf{d} \\ \mathbf{0} & 1 \end{pmatrix}. \tag{2.40}$$

The extended transfer matrix of an accelerator section can be obtained by concatenating the matrices of the individual elements,

$$\mathbf{M}_e = \mathbf{M}_{e,n}\mathbf{M}_{e,n-1}\cdots\mathbf{M}_{e,2}\mathbf{M}_{e,1}. \tag{2.41}$$

From Eq. (2.34), we see that the dispersion vector transports in the same manner as phase space coordinates, except there are perturbations if the accelerator section contains bending fields. This is understandable because the dispersion function is a part of the phase space coordinates. In fact, the dispersion function can be seen as orbit errors due to distributed dipole field errors $1/\rho(s)$. Outside of dipole magnets, there are no orbit error sources and hence the transportation of the dispersion follows $\mathbf{D} = \mathbf{M}\mathbf{D}_0$, which is identical to the transportation of betatron coordinates. Therefore, there exists an invariant of motion similar to J in Eq. (1.63), which is defined by

$$\mathcal{H} = \frac{1}{\beta}\left[D^2 + (\alpha D + \beta D')^2\right]. \tag{2.42}$$

$\mathcal{H}$ is called the dispersion invariant, which is constant in regions without dipole fields. Figure 2.4 shows the dispersion function and the dispersion invariant in the SPEAR3 storage ring.

In a storage ring with a periodic lattice structure, the design dispersion function is also periodic. However, if the bending fields or the linear optics have errors, the dispersion function will deviate from the design. Like the closed

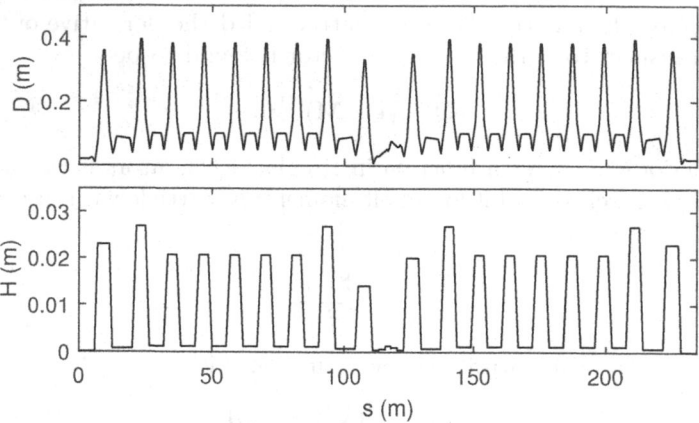

Figure 2.4 The dispersion function (top) and the dispersion invariant (bottom) in the SPEAR3 storage ring.

orbit errors, the errors to the dispersion function are typically dominated by the Fourier harmonics close to the betatron tune.

In a flat accelerator, by design all bending occurs on the horizontal plane. Hence there is no vertical dispersion in the ideal scenario. However, vertical orbit correctors and feed-down effects due to orbit offsets in quadrupole and sextupole magnets give the beam small vertical orbit kicks, which generate vertical dispersion. The horizontal dispersion is also coupled to the vertical plane through skew quadrupoles located at dispersive regions or sextupoles with both vertical orbit offsets and horizontal dispersion. The vertical dispersion generated due to these unintended sources is called the spurious vertical dispersion. In electron storage rings, the spurious vertical dispersion increases the vertical beam emittance.

The path length traveled by an off-momentum particle is closely related to the horizontal dispersion function. The path length over an infinitesimal distance along the reference orbit ds is given by

$$dl = \sqrt{(1 + hx)^2 + x'^2 + y'^2}ds \approx (1 + hx)ds, \qquad (2.43)$$

where $h = \frac{1}{\rho}$ is the curvature of the reference orbit. For an off-momentum particle with momentum deviation δ, from Eq. (2.32), we have $x = x_\beta + D\delta$. The path length difference between an off-momentum particle and the reference particle is

$$\Delta C = \oint (dl - ds) \approx \oint hD\delta ds, \qquad (2.44)$$

where we have dropped the x_β term since it is oscillatory and should largely cancel in the integration. The momentum compaction factor is defined as the derivative of the fractional path length change with respect to the momentum deviation,

$$\alpha_c \equiv \frac{dC}{Cd\delta} = \frac{1}{C} \oint hDds. \qquad (2.45)$$

The momentum compaction factor is very important for the longitudinal motion as it affects the arrival time of off-momentum particles at the RF cavities.

The impact of the dispersion function and the momentum compaction factor to the longitudinal motion is naturally described by the transfer matrix for the 6-dimensional phase space coordinates. To simplify the notation, we omit the vertical plane here. Considering the coordinates $\mathbf{X} = (x, x', z, \delta)$, where $z = -\beta c \Delta t$, the one-turn transfer matrix can be written in the form

$$\mathbf{R} = \begin{pmatrix} \mathbf{M} & \mathbf{0}_{2\times 1} & \mathbf{d} \\ \mathbf{f}^T & 1 & R_{56} \\ \mathbf{0}_{1\times 2} & 0 & 1 \end{pmatrix}, \qquad (2.46)$$

where the subscripts of the $\mathbf{0}$ matrices indicate their dimensions, R_{56} is the $(5, 6)$ element of the full 6-dimensional transfer matrix, which represents the dependence of the z coordinate on the momentum deviation, and $\mathbf{f}$ is a column vector that represents the impact of the horizontal motion on the z-coordinate. Symplecticity of the $\mathbf{R}$-matrix requires that

$$\mathbf{MS_2f} = \mathbf{d}, \qquad (2.47)$$

where $\mathbf{S}_2$ is defined in Eq. (1.45). The momentum compaction factor can be derived from the transfer matrix in Eq. (2.46) by calculating the one-turn shift of the z-coordinate for the off-momentum closed orbit, $(D, D', 1, 0)^T \delta$,

$$\Delta z = -\Delta C = -\alpha_c C \delta = \mathbf{f}^T \mathbf{D} \delta + R_{56} \delta. \qquad (2.48)$$

Using Eq. (2.37) and the Courant-Snyder parametrization of the one-turn transfer matrix $\mathbf{M}$, it can be shown that [53]

$$-\alpha_c C = R_{56} - \mathcal{H} \sin 2\pi \nu_x. \qquad (2.49)$$

Because the dispersion invariant is usually small compared to R_{56}, it is often assumed $-\alpha_c C = R_{56}$.

Eq. (2.46) can also be used to calculate the path length change due to a horizontal corrector kick. The closed orbit shift due to such a corrector kick is given in Eq. (2.8), with which we obtain

$$\Delta z = \mathbf{f}^T \mathbf{X}_c = \mathbf{f}^T (\mathbf{I} - \mathbf{M})^{-1} \begin{pmatrix} 0 \\ \theta \end{pmatrix} = -D\theta, \qquad (2.50)$$

where D is the dispersion at the corrector location. Therefore the path length changes by $\Delta C = D\theta$.

2.4   LINEAR COUPLING

In an ideal lattice that consists of drift spaces, dipoles, and quadrupoles, the beam motion in the horizontal plane is independent of the motion in the vertical plane, and vice versa. However, if skew quadrupole field components are present in the lattice, the motion in the two planes will be coupled. Because magnetic fields in skew quadrupoles are linearly dependent on the transverse coordinates, the coupled beam motion is linear. Main sources of skew quadrupole components are rolls of quadrupole magnets and vertical orbit offsets in sextupole magnets. Solenoid fields also cause linear betatron coupling, although they are not common in high energy accelerators.

Knowing the magnetic fields in a skew quadrupole as given in Eq. (1.19), the equations of motion in a skew quadrupole field are found to be

$$x'' = a_1 y, \qquad y'' = a_1 x, \tag{2.51}$$

where $a_1 = \frac{1}{B\rho}\frac{\partial B_x}{\partial x}$ is the normalized skew quadrupole gradient. The motion in the skew quadrupole field can be solved. In the thin-lens approximation, the solution can be represented by a transfer matrix that relates the coordinate vectors at its entrance and exit faces

$$\mathbf{T}_{sq} = \begin{pmatrix} 1 & 0 & 0 & 0 \\ 0 & 1 & a_1\Delta s & 0 \\ 0 & 0 & 1 & 0 \\ a_1\Delta s & 0 & 0 & 1 \end{pmatrix}, \tag{2.52}$$

where $a_1\Delta s$ is the integrated gradient of the skew quadrupole field. The nonzero elements in the 2×2 off-diagonal blocks couple the (x, x') coordinates and the (y, y') coordinates and hence the motion in the two planes.

The thin-lens skew quadrupole transfer matrix can be written as

$$\mathbf{T}_{sq} = \mathbf{I} + \chi\mathbf{W}_4, \tag{2.53}$$

with the integrated strength $\chi = a_1\Delta s$,

$$\mathbf{W}_4 = \begin{pmatrix} \mathbf{0} & \mathbf{W} \\ \mathbf{W} & \mathbf{0} \end{pmatrix}, \quad \text{and} \quad \mathbf{W} = \begin{pmatrix} 0 & 0 \\ 1 & 0 \end{pmatrix}. \tag{2.54}$$

Suppose a point S is located between points 1 and 2 and the transfer matrix from point 1 to point 2 is $\mathbf{T}_0 = \mathbf{T}_{2S}\mathbf{T}_{S1}$, with

$$\mathbf{T}_{S1} = \begin{pmatrix} \mathbf{M}_1 & \mathbf{0} \\ \mathbf{0} & \mathbf{N}_1 \end{pmatrix}, \quad \mathbf{T}_{2S} = \begin{pmatrix} \mathbf{M}_2 & \mathbf{0} \\ \mathbf{0} & \mathbf{N}_2 \end{pmatrix},$$

when the skew quadrupole is introduced at point S, the new transfer matrix will become

$$\mathbf{T} = \mathbf{T}_{2S}\mathbf{T}_{sq}\mathbf{T}_{S1} = \mathbf{T}_0 + \chi\mathbf{T}_{2S}\mathbf{W}_4\mathbf{T}_{S1}$$

$$= \begin{pmatrix} \mathbf{M}_2\mathbf{M}_1 & \chi\mathbf{M}_2\mathbf{W}\mathbf{N}_1 \\ \chi\mathbf{N}_2\mathbf{W}\mathbf{M}_1 & \mathbf{N}_2\mathbf{N}_1 \end{pmatrix}. \tag{2.55}$$

Using the transfer matrix decomposition in Eq. (1.68) for the unperturbed horizontal and vertical transfer matrices, the off-diagonal blocks of $\mathbf{T}$ can be written as

$$\chi\mathbf{M}_2\mathbf{W}\mathbf{N}_1 = \chi\sqrt{\beta_x\beta_y}\mathbf{B}_{x,2}\mathbf{R}(\psi_{x,2S})\mathbf{W}\mathbf{R}(\psi_{y,S1})\mathbf{B}_{y,1}^{-1},$$

$$\chi\mathbf{N}_2\mathbf{W}\mathbf{M}_1 = \chi\sqrt{\beta_x\beta_y}\mathbf{B}_{y,2}\mathbf{R}(\psi_{y,2S})\mathbf{W}\mathbf{R}(\psi_{x,S1})\mathbf{B}_{x,1}^{-1},$$

where $\beta_{x,y}$ are the beta functions at the skew quadrupole, matrices $\mathbf{B}$ and $\mathbf{R}$ are as defined in Eq. (1.60), and the subscripts indicate the plane and the location. The composition of the upper-right off-diagonal block describes the propagation of the vertical coordinates from point 1 to the skew quadrupole, the application of the horizontal kick to the particle according to its y-coordinate, and the subsequent propagation of the horizontal kick to point 2. Similarly, the lower-left block represents the component of vertical motion at point 2 that comes from the initial horizontal motion at point 1.

When there are multiple skew quadrupoles in the beam line, the transfer matrix can be calculated by concatenating the transfer matrices of all sections and that of the skew quadrupoles. If the coupling strengths, $\chi\sqrt{\beta_x\beta_y}$, of the skew quadrupoles are weak, the higher order terms resulting from the changes of skew quadrupole kicks by the effects of other skew quadrupoles can be neglected. In such cases, the new transfer matrix may be written as the unperturbed transfer matrix plus a summation of the perturbations of all skew quadrupoles in the beam line, i.e.,

$$\mathbf{T} \approx \mathbf{T}_0 + \sum_l \chi_l \mathbf{T}_{2S_l}\mathbf{W}_4\mathbf{T}_{S_l1}, \tag{2.56}$$

where l indicates the l'th skew quadrupole. To first order in χ_l, the skew quadrupoles only change the off-diagonal blocks.

Using the coupled transfer matrix, $\mathbf{T}$, the impact of skew quadrupole components to the beam motion can be described. The non-zero off-diagonal blocks will cause vertical motion for a particle initially launched on the horizontal plane, and vice versa.

In a circular accelerator, the coupled one-turn transfer matrix can be used to track the beam motion for multiple turns. With linear coupling, the horizontal and vertical betatron oscillations will show up in the motion observed on both planes. Figure 2.5 shows the coupled motion observed on the x and y coordinates in a ring for a particle launched with initial offsets of 0.1 mm in both planes. The Fourier spectrum of the x motion includes a component with the vertical tune, while the y spectrum includes the horizontal tune.

Linear coupling also causes the closed orbit to be coupled between the two transverse planes. With linear coupling, a kick on the horizontal plane causes orbit deviations not only in the horizontal plane, but also in the vertical plane. Similarly, a vertical kick causes both horizontal and vertical orbit deviations. The orbit deviations in the other plane are small if the coupling is weak.

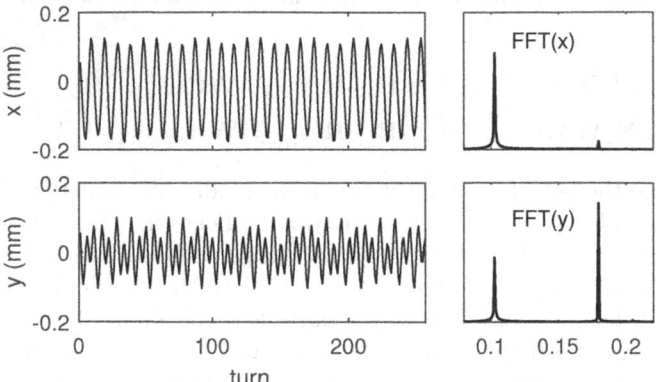

Figure 2.5 The coupled motion in the SPEAR3 storage ring for a particle launched with initial offsets of $x = y = 0.1$ mm. The coupling is introduced by 15 skew quadrupoles with integrated strengths randomly chosen from a Gaussian distribution with $\sigma_\chi = 0.012$ m^{-1}. The uncoupled betatron tunes are $\nu_x = 14.106$ and $\nu_y = 6.177$.

As shown in Figure 2.5, the coupled motion in a ring consists of two frequency components in both transverse planes. These two frequency components correspond to two independent modes, which are called decoupled modes or normal modes. The normal mode coordinates, $\hat{\mathbf{X}} = (u_a, u_a', u_b, u_b')^T$ are related to the usual coordinates $\mathbf{X} = (x, x', y, y')^T$ through a linear transformation,

$$\hat{\mathbf{X}} = \mathbf{V}^{-1}\mathbf{X}, \tag{2.57}$$

where $\mathbf{V}$ is a 4×4 symplectic matrix. The one-turn transfer matrix for the normal coordinates, $\hat{\mathbf{X}}$, is block diagonal, i.e.,

$$\hat{\mathbf{T}} = \mathbf{V}^{-1}\mathbf{T}\mathbf{V} = \begin{pmatrix} \mathbf{M}_a & \mathbf{0} \\ \mathbf{0} & \mathbf{M}_b \end{pmatrix}. \tag{2.58}$$

The one-turn transfer matrices for mode a and b, $\mathbf{M}_a$ and $\mathbf{M}_b$, respectively, can be Courant-Snyder parametrized.

It has been shown that for a general 4×4 coupled transfer matrix [105]

$$\mathbf{T} = \begin{pmatrix} \mathbf{M} & \mathbf{m} \\ \mathbf{n} & \mathbf{N} \end{pmatrix}, \tag{2.59}$$

the linear transformation matrices to and from the decoupled coordinates are given by

$$\mathbf{V} = \begin{pmatrix} r\mathbf{I} & \mathbf{C} \\ -\mathbf{C}^+ & r\mathbf{I} \end{pmatrix}, \qquad \mathbf{V}^{-1} = \begin{pmatrix} r\mathbf{I} & -\mathbf{C} \\ \mathbf{C}^+ & r\mathbf{I} \end{pmatrix}, \tag{2.60}$$

where $\mathbf{C}^+$ is the symplectic conjugate of $\mathbf{C}$, defined as

$$\mathbf{C}^+ \equiv \mathbf{S}_2^T \mathbf{C}^T \mathbf{S}_2 = \begin{pmatrix} C_{22} & -C_{12} \\ -C_{21} & C_{11} \end{pmatrix}, \tag{2.61}$$

and the matrix $\mathbf{C}$ is defined as

$$\mathbf{C} = -\frac{\mathbf{H}\mathrm{sgn}[\mathrm{Tr}(\mathbf{M} - \mathbf{N})]}{r\sqrt{[\mathrm{Tr}(\mathbf{M} - \mathbf{N})]^2 + 4\|\mathbf{H}\|}}, \tag{2.62}$$

with $\mathbf{H} \equiv \mathbf{m} + \mathbf{n}^+$, $\|\mathbf{H}\|$ is the determinant of $\mathbf{H}$, $\mathrm{sgn}(\cdot)$ gives the sign, and

$$r = \sqrt{\frac{1}{2} + \frac{1}{2}\sqrt{\frac{[\mathrm{Tr}(\mathbf{M} - \mathbf{N})]^2}{[\mathrm{Tr}(\mathbf{M} - \mathbf{N})]^2 + 4\|\mathbf{H}\|}}}, \tag{2.63}$$

respectively. Parameter r and the determinant of the matrix $\mathbf{C}$ satisfy

$$r^2 + \|\mathbf{C}\| = 1. \tag{2.64}$$

Eqs. (2.59-2.63) give a procedure to decouple the linearly coupled motion between two the planes. Knowing the $\mathbf{V}$ matrix, the usual phase space coordinates can be expressed in terms of the decoupled coordinates,

$$\begin{pmatrix} x \\ x' \end{pmatrix} = r \begin{pmatrix} u_a \\ u_a' \end{pmatrix} + \mathbf{C} \begin{pmatrix} u_b \\ u_b' \end{pmatrix}, \quad \begin{pmatrix} y \\ y' \end{pmatrix} = -\mathbf{C}^+ \begin{pmatrix} u_a \\ u_a' \end{pmatrix} + r \begin{pmatrix} u_b \\ u_b' \end{pmatrix}. \tag{2.65}$$

Clearly, the motion in the x or y plane contains the components of both normal modes. If we can separate the normal mode components in the turn-by-turn motion observed in the two planes, we can obtain information about the decoupling matrix $\mathbf{C}$, which in turn can be used to derive information about the coupling sources.

Conversely, the transformation also allows us to calculate the projection of the motion given in the x and y planes onto the two normal modes. This can be used to calculate the excitation of the normal modes by, for example, photon emissions. When a photon is emitted, the betatron coordinates change by $\Delta x_\beta = -u D_x$ and $\Delta x_\beta' = -u D_x'$, where u is the fractional momentum loss of the particle due to the photon emission. The changes of the normal mode coordinates will be

$$\begin{pmatrix} \Delta u_a \\ \Delta u_a' \end{pmatrix} = (-u) r \begin{pmatrix} \Delta D \\ \Delta D' \end{pmatrix}, \quad \begin{pmatrix} \Delta u_b \\ \Delta u_b' \end{pmatrix} = (-u) \mathbf{C}^+ \begin{pmatrix} \Delta D \\ \Delta D' \end{pmatrix}. \tag{2.66}$$

The increment of the b-mode (vertical) action is

$$\Delta J_b = \frac{1}{2\beta_b} \left(\Delta u_b^2 + (\alpha_b \Delta u_b + \beta_b \Delta u_b')^2 \right),$$

which can then be used to calculate the betatron coupling contribution to the equilibrium vertical emittance in electron storage rings.

The elements of the decoupling matrix $\mathbf{C}$ can be related to the distribution of the coupling sources. To that end we first need to calculate matrix $\mathbf{H}$. For simplicity, we define matrix

$$\bar{\mathbf{H}} = \mathbf{B}_x^{-1}\mathbf{H}\mathbf{B}_y, \tag{2.67}$$

which is the equivalent of $\mathbf{H}$ for the normalized coordinates $(\bar{x}, \bar{x}', \bar{y}, \bar{y}')$ (see Eq. (1.64)). From Eq. (2.56), it can be shown that if we define

$$h_- = 2i \sin \pi(\nu_x + \nu_y)e^{i\pi(\nu_x-\nu_y)} \sum_l \chi_l \sqrt{\beta_{x,l}\beta_{y,l}}e^{-i(\Psi_{x,l}-\Psi_{y,l})}, \tag{2.68}$$

$$h_+ = 2i \sin \pi(\nu_x - \nu_y)e^{i\pi(\nu_x+\nu_y)} \sum_l \chi_l \sqrt{\beta_{x,l}\beta_{y,l}}e^{i(\Psi_{x,l}+\Psi_{y,l})}, \tag{2.69}$$

the elements of $\bar{\mathbf{H}}$ are given by

$$\bar{H}_{11} = \frac{1}{2}\mathrm{Im}(h_- + h_+), \quad \bar{H}_{12} = \frac{1}{2}\mathrm{Re}(h_- - h_+), \tag{2.70a}$$

$$\bar{H}_{21} = -\frac{1}{2}\mathrm{Re}(h_- + h_+), \quad \bar{H}_{22} = \frac{1}{2}\mathrm{Im}(h_- - h_+). \tag{2.70b}$$

The determinant of $\mathbf{H}$ is thus (since $||\mathbf{B}_{x,y}|| = 1$)

$$||\mathbf{H}|| = ||\bar{\mathbf{H}}|| = \frac{1}{4}(|h_-|^2 - |h_+|^2). \tag{2.71}$$

Defining the coupling coefficients

$$G_\pm = \frac{1}{2\pi} \sum_l \chi_l \sqrt{\beta_{x,l}\beta_{y,l}}e^{\pm i(\Psi_{x,l}\pm\Psi_{y,l})}, \tag{2.72}$$

we have

$$r^2 = \frac{1}{2} + \frac{1}{2}\left(1 + \frac{\pi^2|G_-|^2}{\sin^2 \pi(\nu_x - \nu_y)} - \frac{\pi^2|G_+|^2}{\sin^2 \pi(\nu_x + \nu_y)}\right)^{-\frac{1}{2}}. \tag{2.73}$$

Eq. (2.73) indicates that if the betatron tunes satisfy the linear difference resonance condition, i.e., $\nu_x - \nu_y = p$, where p is an integer, the G_- term dominates and $r^2 = ||\mathbf{C}|| = \frac{1}{2}$. In this case the excitation of motion is equally split between the two normal modes. If the tunes are near the linear sum resonance, with $\nu_x+\nu_y \approx p$, the G_+ term dominates. As $|\nu_x+\nu_y-p|$ approaches $|G_+|$, r^2 and $\mathrm{Abs}(||\mathbf{C}||)$ tend to infinity, where $\mathrm{Abs}(\cdot)$ stands for the absolute value. The beam motion near the linear sum resonance is unstable.

In a typical storage ring, the betatron tunes are closer to the linear difference resonance than the linear sum resonance, i.e., the fractional parts of ν_x and ν_y tend to be both in the $[0, 0.5]$ zone, or the $[0.5, 1]$ zone. The linear coupling is usually weak, with

$$|G_-| \ll |\nu_x - \nu_y - p|, \tag{2.74}$$

satisfied and, correspondingly, $\text{Abs}(||\mathbf{C}||) \ll r^2 \approx 1$. However, for the next generation storage ring light sources with the natural emittance at the diffraction limit, it is desirable to operate near or on the linear difference resonance. In this case, $|\nu_x - \nu_y - p| \leq |G_-|$, and hence the horizontal and vertical emittances are nearly equal. The beam in this condition is called a round beam.

2.5   CHROMATIC EFFECT

As discussed in Section 2.3, the energy dependence of the bending angles by dipole fields gives rise to dispersion, the orbit dependence on beam energy. As the multipole field one order higher than the dipole, quadrupole fields determine the linear optics of the lattice. The focusing strength of a quadrupole also depends on the beam energy. The dependence of the linear optics on the beam energy is called the chromatic effect.

From Eq. (1.36) we see that the focusing gradient for an off-momentum particle is $K = \frac{b_1}{1+\delta}$, where $b_1 = \frac{1}{B\rho}\frac{\partial B_y}{\partial x}$ is the focusing gradient for the reference particle, and δ is the momentum deviation of the off-momentum particle. Therefore, the transfer matrix of the quadrupole for the off-momentum particle is different from that of the reference particle. In a circular accelerator, the one-turn transfer matrix depends on the momentum deviation, which means the betatron tunes, betatron phase advances, and Courant-Snyder parameters all depend on the beam energy.

The focusing error of a quadrupole for an off-momentum particle is

$$\Delta K_x = -(b_1 + 2h^2)\delta \approx -K_x\delta, \quad \Delta K_y = b_1\delta = -K_y\delta, \qquad (2.75)$$

where $K_x = b_1 + h^2$, $K_y = -b_1$ are the horizontal and vertical focusing functions, respectively, and h is the curvature of the reference orbit. The focusing errors for off-momentum particles affect the linear optics of the off-momentum particles in the same manner as the quadrupole errors we previously studied. For example, the betatron tune of a ring would be changed by

$$\Delta\nu \approx \frac{1}{4\pi} \oint \beta(-K\delta)ds, \qquad (2.76)$$

where $-K\delta$ is the error of the focusing function for a particle with momentum deviation δ. The derivative of the momentum dependent tune shift with respect to the momentum deviation is called the chromaticity,

$$C \equiv \frac{d\nu}{d\delta}. \qquad (2.77)$$

The uncorrected chromaticity is called natural chromaticity,

$$C_{\text{nat}} \approx -\frac{1}{4\pi} \oint K\beta ds. \qquad (2.78)$$

The natural chromaticity is a negative quantity because a particle with higher momentum receives less focusing by the quadrupoles.

For example, the horizontal transfer matrix for a FODO cell for an off-momentum particle would need to change from Eq. (1.50) by replacing the focal length, f, with $f(1 + \delta)$. The betatron tune contribution of the cell will change accordingly. From Eqs. (1.65) and (2.77), the natural chromaticity of a FODO cell is found to be

$$C^{\text{FODO}} = -\nu \frac{\tan \Phi/2}{\Phi/2} \approx -\nu, \tag{2.79}$$

where the approximation is valid when the phase advance on the cell is small.

High intensity bunched beams in a storage ring can suffer from the head-tail instability if the ring has a large, negative chromaticity (if above transition). The large tune spread in the beam due to a large chromaticity can also cause difficulties. The correction of the natural chromaticity is usually necessary for storage rings. Typically the horizontal and vertical chromaticities are corrected to slightly positive numbers (above transition).

Chromaticity correction is achieved by placing sextupole magnets in the lattice at dispersive locations. At such locations an off-energy particle travels on an orbit with a horizontal offset from the magnet center. On this orbit the particle sees a quadrupole field component from the feed-down effect (see the illustration in Figure 2.6). For a particle with momentum deviation δ which is on the dispersion orbit of $\Delta x = D\delta$ at the sextupole location, the quadrupole component it sees is

$$\Delta K(\delta) = b_2 D\delta, \tag{2.80}$$

where b_2 is the normalized sextupole strength. This momentum dependent quadrupole error affects the chromaticities. Including the corrections, the horizontal and vertical chromaticities are

$$C_x \approx -\frac{1}{4\pi} \oint \beta_x (K_x - b_2 D)ds, \tag{2.81a}$$

$$C_y \approx -\frac{1}{4\pi} \oint \beta_y (K_y + b_2 D)ds, \tag{2.81b}$$

respectively. Because the momentum dependent quadrupole error is focusing in one transverse plane and defocusing in the the other plane, at least two sextupole families are required for chromaticity correction in both planes. One sextupole family is located where $\beta_x > \beta_y$, while the other sextupole family is located where $\beta_y < \beta_x$. A large disparity in the horizontal and vertical beta functions at the sextupole locations helps reduce the required sextupole strengths for chromaticity correction.

The focusing errors for off-momentum particles by the quadrupole magnets not only change the off-momentum betatron tunes, but also the off-momentum beta functions and phase advances. The beta beating for off-momentum particles is called chromatic beta beating. A large chromatic beta beating could lead to a reduced momentum acceptance. Since the focusing errors cannot

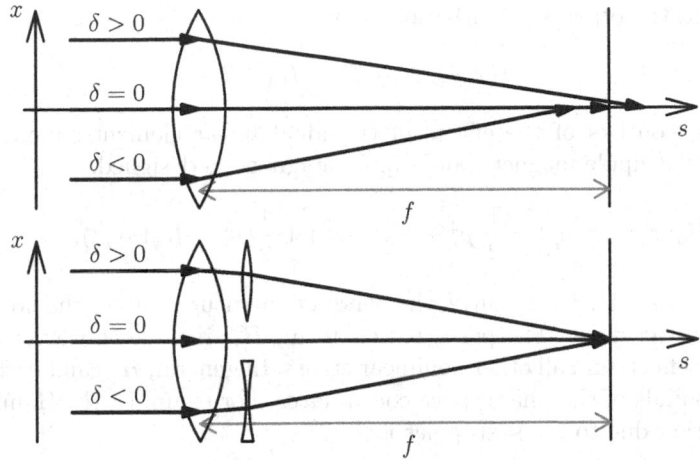

Figure 2.6 Illustration of chromaticity correction with sextupole magnets. Top: focusing error of a quadrupole for off-momentum particles. Bottom: the feed-down quadrupole field from a sextupole magnet provides correction to the focusing errors.

be corrected exactly at the locations of the error sources, correction of the chromaticities usually does not completely eliminate chromatic beta beating. The chromatic beta beating due to quadrupoles and sextupoles in the lattice is a systematic error. It is important to properly arrange these magnets in the lattice design in order to avoid excessive chromatic beta beating.

2.6 NONLINEAR BEAM DYNAMICS

Sextupole magnets are introduced into circular accelerators to correct chromaticities. The magnetic fields in sextupoles are nonlinear with respect to transverse positions of the beam particles. The nonlinear forces can lead to unstable beam motion and beam loss when the transverse offsets are sufficiently large as the particle motion can be driven onto nonlinear resonances and become unstable under large oscillation amplitudes. A large stability region is critical for storage ring lattice designs.

The nonlinear beam motion in circular accelerator lattices can be analyzed with the Hamiltonian dynamics approach or the Lie map approach.

2.6.1 Hamiltonian dynamics approach

In general, the Hamiltonian that describes the beam motion in an accelerator lattice can be split into two parts, one representing the ideal linear motion by design and the other a perturbation term including all other effects,

$$H(x, p_x, y, p_y, \delta) = H_0 + H_1, \tag{2.82}$$

where H_0 consists of the effects of the ideal linear elements, namely, drift spaces, quadrupole magnets, and dipole magnets as designed,

$$H_0(x, p_x, y, p_y) = \frac{1}{2}(p_x^2 + K_x(s)x^2) + \frac{1}{2}(p_y^2 + K_y(s)y^2), \tag{2.83}$$

with $K_x = b_1 + h^2$, $K_y = -b_1$, h the bending curvature, and b_1 the normalized quadrupole gradient. The perturbation term, H_1, includes linear errors, the sextupole effect, and all other nonlinear errors. In general, H_1 can be expressed in polynomials of the phase space coordinates. For example, the Hamiltonian perturbation due to the sextupoles is

$$H_{1,\text{sext}} = b_2(s)\frac{x^3 - 3xy^2}{6}. \tag{2.84}$$

The linear terms in a sextupole Hamiltonian are the same as a drift space and are included in H_0.

The linear motion represented by H_0 can be described by 4×4 transfer matrices. By introducing a few canonical coordinate transformations, the stable linear beam motion can be cast into two uncoupled harmonic oscillations. In the action-angle coordinates of the oscillations, the new Hamiltonian becomes

$$\tilde{H}_0 = RH_0 = v_x J_x + v_y J_y, \tag{2.85}$$

with

$$
\begin{aligned}
x &= x_\beta + D\delta, \quad p_x = p_{x\beta} + D'\delta, \\
x_\beta &= \sqrt{2\beta_x J_x} \cos \Phi_x, \quad \beta_x p_{x\beta} + \alpha_x x_\beta = -\sqrt{2\beta_x J_x} \sin \Phi_x, \\
y &= \sqrt{2\beta_y J_y} \cos \Phi_y, \quad \beta_y p_y + \alpha_y y = -\sqrt{2\beta_y J_y} \sin \Phi_y,
\end{aligned}
\tag{2.86}
$$

where $\Phi_{xy} = \phi_{xy} + \psi_{xy} - v_{xy}\theta$, $(\phi_{x,y}, J_{x,y})$ are action-angle coordinate pairs, $\psi_{x,y}$ are betatron phase advances, and $\theta = s/R$ is used as the free variable (with $2\pi R$ the ring circumference). The factor of R in the new Hamiltonian comes from the change of free variable from s to θ. If we further introduce coordinates

$$
\begin{aligned}
z_1 &= \sqrt{2\beta_x J_x} e^{i\Phi_x}, \quad \bar{z}_1 = \sqrt{2\beta_x J_x} e^{-i\Phi_x}, \\
z_2 &= \sqrt{2\beta_y J_y} e^{i\Phi_y}, \quad \bar{z}_2 = \sqrt{2\beta_y J_y} e^{-i\Phi_y},
\end{aligned}
$$

the perturbation Hamiltonian can be rewritten as

$$\tilde{H}_1 = R \sum_{jklmn \geq 0} H_{jklmn}(\theta) z_1^j \bar{z}_1^k z_2^l \bar{z}_2^m \delta^n,$$

$$= R \sum H_{jklmn}(\theta) \beta_x^{\frac{j+k}{2}} \beta_y^{\frac{l+m}{2}} J_x^{\frac{j+k}{2}} J_y^{\frac{l+m}{2}} \delta^n$$

$$e^{i[(j-k)(\psi_x - \nu_x \theta) + (l-m)(\psi_y - \nu_y \theta)]} e^{i[(j-k)\phi_x + (l-m)\phi_y]},$$

$$\equiv \sum h_{jklmn}(\theta) J_x^{\frac{j+k}{2}} J_y^{\frac{l+m}{2}} e^{i[(j-k)\phi_x + (l-m)\phi_y]} \delta^n, \qquad (2.87)$$

where

$$h_{jklmn}(\theta) = R H_{jklmn}(\theta) \beta_x^{\frac{j+k}{2}} \beta_y^{\frac{l+m}{2}} e^{i[(j-k)(\psi_x - \nu_x \theta) + (l-m)(\psi_y - \nu_y \theta)]}. \qquad (2.88)$$

The Hamiltonian $\tilde{H}_1$ is said to be given in the resonance basis.

The functions $h_{jklmn}(\theta)$ are periodic with respect to θ with the period of 2π and can be Fourier expanded,

$$h_{jklmn}(\theta) = \sum_{p=-\infty}^{\infty} h_{jklmn}^{(p)} e^{ip\theta}, \qquad (2.89)$$

with the Fourier coefficients given by

$$h_{jklmn}^{(p)} = \frac{1}{2\pi} \oint d\theta \, h_{jklmn}(\theta) e^{-ip\theta}. \qquad (2.90)$$

With the Fourier expansion, the perturbation Hamiltonian is now written as

$$\tilde{H}_1 = \sum_{jklmn \geq 0} \sum_{p=-\infty}^{\infty} h_{jklmn}^{(p)} J_x^{\frac{j+k}{2}} J_y^{\frac{l+m}{2}} e^{i[(j-k)\phi_x + (l-m)\phi_y + p\theta]} \delta^n. \qquad (2.91)$$

The various terms in the perturbation Hamiltonian have different impact on the beam motion. The terms with $n > 0$ affect off-momentum particles and can be referred to as chromatic terms, while the $n = 0$ terms affect the on-momentum particle motion and are called geometric terms. The terms with $j = k$ and $l = m$ are independent of the angle coordinates. Among these terms, the ones with $p = 0$ cause the tunes to change, which can be seen from the Hamilton's equation

$$\Delta \nu_x = \frac{\partial \tilde{H}_1}{\partial J_x}, \qquad \Delta \nu_y = \frac{\partial \tilde{H}_1}{\partial J_y}. \qquad (2.92)$$

For example, the terms $h_{11000}^{(0)}$ and $h_{00110}^{(0)}$ correspond to betatron tune changes due to quadrupole errors around the ring,

$$\Delta \nu_x = h_{11000}^{(0)}, \qquad \Delta \nu_y = h_{00110}^{(0)}. \qquad (2.93)$$

The terms $h_{11001}^{(0)}$ and $h_{00111}^{(0)}$ correspond to tune changes for off-momentum particles, namely, the chromaticities.

$$\Delta C_x = h_{11001}^{(0)}, \qquad C_y = h_{00111}^{(0)}. \tag{2.94}$$

The tune changes due to the above terms are independent of the action variables. For most terms, the tune shifts depend on the action variables, or equivalently, the oscillation amplitudes; such effects are called tune shifts with amplitude, amplitude-dependent detuning, or nonlinear detuning. The leading terms in the perturbation Hamiltonian that give rise to nonlinear detuning are

$$\Delta \tilde{H}_1 = h_{22000}^{(0)} J_x^2 + h_{11110}^{(0)} J_x J_y + h_{00220}^{(0)} J_y^2. \tag{2.95}$$

The tune shifts with amplitude from these terms can be characterized by the following coefficients.

$$\frac{\partial \nu_x}{\partial J_x} = 2h_{22000}^{(0)}, \qquad \frac{\partial \nu_x}{\partial J_y} = \frac{\partial \nu_y}{\partial J_x} = h_{11110}^{(0)}, \qquad \frac{\partial \nu_y}{\partial J_y} = 2h_{00220}^{(0)}. \tag{2.96}$$

There are also higher order terms that cause tune shifts dependence on the action variables to higher orders.

The terms in Eq. (2.91) that have either $j \neq k$ or $l \neq m$ or both depend on the angle coordinates ϕ_x or ϕ_y. Most of these terms have little impact over the motion of the particles because they oscillate quickly with time and hence the average effect is negligible. However, for some terms, the phase factor may be slowly varying. The effect from these terms can build up and fundamentally change the behavior of particle motion. These terms satisfy the resonance condition

$$(j-k)\frac{d\phi_x}{d\theta} + (l-m)\frac{d\phi_y}{d\theta} + p = (j-k)\nu_x + (l-m)\nu_y + p,$$

$$= n_1\nu_x + n_2\nu_y + p \approx 0, \tag{2.97}$$

where $n_1 = j - k$ and $n_2 = l - m$. The order of the resonance is defined as $|n_1| + |n_2|$. Beam motion around nonlinear resonances can become unstable, as the resonances can drive the particles to large oscillation amplitudes and cause beam loss. To alleviate the impact of the nonlinear resonances, it is desirable to minimize the strengths of the resonance harmonics, $h_{jklmn}^{(p)}$.

Periodicity in a lattice can automatically set many systematic (i.e., inherent in the design) resonance terms to zero. If a ring consists of N repetitive, identical cells, the resonance harmonics reduce to

$$h_{jklm}^{(p)} = \frac{1}{2\pi} \left(\int_{\text{cell}} h_{jklm} e^{-jp\theta} d\theta \right) \sum_{q=0}^{N-1} e^{-i2\pi q \frac{p}{N}},$$

$$= \frac{1}{2\pi} \left(\int_{\text{cell}} h_{jklm} e^{-jp\theta} d\theta \right) e^{-i\pi p \frac{N-1}{N}} \frac{\sin p\pi}{\sin(p\pi/N)}, \tag{2.98}$$

which vanish, unless p is either 0 or a multiple of N. Therefore, it is beneficial to retain high periodicity in the design of a storage ring lattice.

In a typical storage ring, the main sources of nonlinearity in the beam motion are the sextupole magnets. From Eq. (2.84), it is straightforward to show that in the resonance basis the sextupole Hamiltonian consists of the following geometric terms: h_{3000}, h_{0300}, h_{2100}, h_{1200}, h_{1011}, h_{0111}, h_{1020}, h_{0102}, h_{1002}, and h_{0120} (here the fifth index, 0, is suppressed in the subscript). Each of these terms contributes to driving a corresponding resonance. These resonances include $3\nu_x = p$, $\nu_x = p$, $\nu_x + 2\nu_y = p$, and $\nu_x - 2\nu_y = p$. The above resonances are driven by sextupoles through their direct impact on the linear motion, which corresponds to the first order perturbation to the linear motion. The effect of sextupole fields on the nonlinear beam motion perturbed by other sextupoles or themselves on previous passes gives rise to additional resonances, which correspond to terms from the second or higher order perturbations. Resonances driven by sextupoles through the second order perturbation include $4\nu_x = p$, $2\nu_x = p$, $2\nu_y = p$, $4\nu_x \pm 2\nu_y = p$, and $2\nu_x \pm 2\nu_y = p$.

Aside from the systematic nonlinear resonances driven by sextupoles in the design, a lattice always has field errors that are systematic or random deviations from the design. The field errors will drive many nonlinear resonances. When the tunes of oscillating particles are shifted onto certain nonlinear resonances due to nonlinear detuning or linear and nonlinear chromaticities in a storage ring, the particles can get lost. Particle loss from the nonlinear beam motion limits the dynamic aperture and the local momentum aperture. Sufficiently large dynamic aperture and local momentum aperture are basic requirements for the operation of a storage ring. The lattice design of a storage ring often relies on extensive optimization of the linear and nonlinear optics to achieve the desired nonlinear dynamics performance. However, linear and nonlinear errors in the real machine cause the operation conditions to deviate from the design. During the commissioning phase, it is necessary to correct the errors in the machine in order to restore the lattice performance. Beam-based optimization may be used to compensate the effects of the errors when direct correction methods are not available.

2.6.2 Lie map approach

The Hamiltonian dynamics approach gives a continuous description of the beam motion. In the accelerator context, it often suffices to know the transfer map between two locations. The transfer map can be given by a Taylor expansion of the coordinates at location 2 in terms of the coordinates at location 1, as shown in Eq. (1.48). The Taylor map is easy to evaluate, but it quickly becomes large and cumbersome when it is extended to higher orders. More importantly, the truncated Taylor map is generally not symplectic and hence not ideal for the study of long term stability of beam motion. The Lie map is an alternative representation of the transfer map, which is not only symplectic but also compact.

In a Hamiltonian system, the time derivative of a function of the phase space coordinates, $f(\mathbf{X})$, is given by

$$f' = \frac{\partial f}{\partial s} + \sum_i [\frac{\partial f}{\partial x_i} x'_i + \frac{\partial f}{\partial p_i} p'_i] = \frac{\partial f}{\partial s} + [f, H], \qquad (2.99)$$

where $'$ represents derivative with respect to the free variable, s, summation is over pairs of conjugate coordinates, $H(\mathbf{X}; s)$ is the Hamiltonian, and the Poisson bracket for functions f and g is defined as

$$[f, g] = \sum_i \frac{\partial f}{\partial x_i} \frac{\partial g}{\partial p_i} - \frac{\partial f}{\partial p_i} \frac{\partial g}{\partial x_i} \equiv: f : g, \qquad (2.100)$$

where $: f : g$ is simply another notation for the Poisson bracket. Assuming no explicit time dependence in both f and H, the higher order time derivatives of f can be readily obtained

$$f' = : -H : f, \quad f'' = : (-H)^2 : f, \quad \cdots, \quad f^{(n)} = : (-H)^n : f, \qquad (2.101)$$

where $-H$ is used because $[f, H] = [-H, f]$. For an accelerator element with constant magnetic field profile over length L, the function f at the exit face can be expressed in a Taylor series of coordinates at the entrance face

$$f(\mathbf{X}_2) = f_1 + f'_1 L + \frac{1}{2} f''_1 L^2 + \cdots = e^{:-HL:} f(\mathbf{X})|_{\mathbf{X}=\mathbf{X}_1}, \qquad (2.102)$$

where subscript 1 indicates values at the entrance face and $e^{:g:}$ is defined as a Lie map with the generating function g,

$$e^{:g:} \equiv 1 + : g : + \frac{1}{2} : g :^2 + \cdots \frac{1}{n!} : g :^n + \cdots. \qquad (2.103)$$

When Eq. (2.102) is applied to the phase space coordinates, $\mathbf{X}$, it gives the transfer map. Hence, for a typical accelerator element the transfer map is a Lie map with generating function $g = -H(\mathbf{X})L$. For linear optics elements, the generating functions are quadratic functions of the phase space coordinates. For example, the Lie map for a drift space is $\exp(: -\frac{1}{2}(p_x^2 + p_y^2)L :)$. The map for the linear elements can be expressed in closed forms. However, the maps for nonlinear elements, such as a sextupole,

$$f_{\text{sext}} = -(\frac{p_x^2 + p_y^2}{2} + \frac{K_2}{6}(x^3 - 3xy^2))L, \qquad (2.104)$$

do not have closed forms.

When two elements are joined together, the Lie map for the section consisting the two elements can be obtained by concatenating the two individual Lie maps using the Baker-Campbell-Hausdorff (BCH) formula,

$$e^{:f:} e^{:g:} = e^{:h:}, \qquad (2.105)$$

where f and g are the generating functions for the first and second elements (noting the order), respectively, and

$$h = f + g + \frac{1}{2} : f : g + \frac{1}{12} : f :^2 g + \frac{1}{12} : g :^2 f + \cdots . \qquad (2.106)$$

With concatenation, the Lie map for any section with multiple elements can be obtained. The one-turn map at a location in a ring is a special example.

The terms in the generating function can be grouped by their orders in the polynomial,

$$e^{:h:} = e^{:f_2:} e^{:f_3:} e^{:f_4:} \cdots , \qquad (2.107)$$

where f_2 contains all quadratic terms, f_3 all third order terms, etc. Note that f_3, f_4, and higher order terms differ from terms in h because extra terms are generated when the BCH formula is used to separate the map in Eq. (2.107). The quadratic terms in f_2 are the same as in h and can be expressed in terms of the Courant-Snyder parameters,

$$f_2 = -\frac{\pi\nu_x}{\beta_x}(x^2 + (\alpha_x x + \beta p_x)^2) - \frac{\pi\nu_y}{\beta_y}(y^2 + (\alpha_y y + \beta p_y)^2). \qquad (2.108)$$

The f_2 map represents the linear motion, while the f_3, f_4, and higher order terms give rise to nonlinear detuning and resonances.

The Lie map can be brought into a simple form called the normal form [31, 11, 12] through a coordinate transformation. As a first step, the resonance basis coordinates are introduced, using the normalized coordinates defined in Eq. (1.64),

$$h_x^\pm = \bar{x} \pm i\bar{p}_x = \sqrt{2J_x}e^{\mp i\phi_x}, \qquad h_y^\pm = \bar{y} \pm i\bar{p}_y = \sqrt{2J_y}e^{\mp i\phi_x}, \qquad (2.109)$$

where (J_x, ϕ_x) and (J_y, ϕ_x) are action-angle variables for the two transverse planes, respectively. Beside the terms that describe the ideal linear motion, the remainder of the generating function, including linear errors and all nonlinear terms, can be expressed in the resonance basis

$$\Delta h = \sum_{jklm} h_{jklm} h_x^j h_x^{-k} h_y^l h_y^{-m}. \qquad (2.110)$$

The terms with both $j = k$ and $l = m$ do not involve the angle coordinates and hence will not cause variations to the action variables. Instead, they will change the betatron tunes. The terms with either $j \neq k$ or $l \neq m$ will drive resonances and are referred to as resonance driving terms (RDTs).

The coordinate transformation to the normal form coordinates can be cast into a Lie map,

$$\zeta = e^{-:F:}h, \quad \text{with } F = \sum_{jklm} f_{jklm} \zeta_x^j \zeta_x^{-k} \zeta_y^l \zeta_y^{-m}, \qquad (2.111)$$

where $\mathbf{h} = (h_x^+, h_x^-, h_y^+, h_y^-)^T$ are the original coordinates, $\boldsymbol{\zeta} = (\zeta_x^+, \zeta_x^-, \zeta_y^+, \zeta_y^-)^T$ are the normal form coordinates, and

$$\zeta_{x,y}^{\pm} = \sqrt{2I_{x,y}}e^{\mp i\psi_{x,y}}, \tag{2.112}$$

with new action-angle coordinates (I_x, ψ_x) and (I_y, ψ_y). The generating function, F, is chosen to make the motion in the new coordinate as simple as possible. In the non-resonant case, the beam motion in the new resonance basis will be a simple rotation. It can be shown that the coefficients, f_{jklm} and h_{jklm}, are connected through [36]

$$f_{jklm} = \frac{h_{jklm}}{1 - e^{i2\pi[(j-k)\nu_x + (l-m)\nu_y]}}, \tag{2.113}$$

where only the resonance driving terms ($j \neq k$ or $l \neq m$ or both) are kept.

The beam motion in the original coordinates can be determined from the inverse coordinate transformation, i.e., $e^{:F:}\zeta_x^-$, hence [10]

$$h_x^- \approx \zeta_x^- + [F, \zeta_x^-] = \zeta_x^- - 2i \sum_{jklm} jf_{jklm}(\zeta_x^+)^{j-1}\zeta_x^{-k}(\zeta_y^+)^l\zeta_y^{-m}, \tag{2.114}$$

and similarly for the vertical plane. The new resonance basis coordinates after N turns will be

$$\zeta_{x,y}^-(N) = \sqrt{2I_{x,y}}e^{i(2\pi\nu_{x,y}N + \psi_{x0,y0})}, \tag{2.115}$$

where the tunes may include any nonlinear detuning. Therefore, the original resonance basis coordinates after N turns are given by [10]

$$h_x^-(N) = \sqrt{2I_x}e^{i(2\pi\nu_x + \psi_{x0})} - 2i \sum_{jklm} jf_{jklm}(2I_x)^{\frac{j+k-1}{2}}(2I_y)^{\frac{l+m}{2}}$$
$$\cdot e^{i[(1-j+k)(2\pi\nu_x N + \psi_{x0}) + (m-l)(2\pi\nu_y N + \psi_{y0})]}, \tag{2.116}$$
$$h_y^-(N) = \sqrt{2I_y}e^{i(2\pi\nu_y + \psi_{y0})} - 2i \sum_{jklm} lf_{jklm}(2I_x)^{\frac{j+k}{2}}(2I_y)^{\frac{l+m-1}{2}}$$
$$\cdot e^{i[(k-j)(2\pi\nu_x N + \psi_{x0}) + (1-l+m)(2\pi\nu_y N + \psi_{y0})]}. \tag{2.117}$$

With Eqs. (2.110), (2.113), and (2.116)-(2.117), the observed beam motion is related to the RDTs in the generating function of the one-turn Lie map. Each term in Eqs. (2.116)-(2.117) corresponds to a spectral line on the turn-by-turn orbit data, while a spectral line typically has contributions from many terms. For example, the third order RDT h_{3000}, which is proportional to sextupole strengths, drives the resonance $3\nu_x = p$ and its corresponding spectral line on the horizontal turn-by-turn orbit data is $1 - 2\nu_x$.

2.7 LATTICE MODELING AND PARTICLE TRACKING

While beam dynamics theories are very useful for understanding the nature of the beam motion, the design and operation of an accelerator often demand more precise and more detailed description of the beam motion and beam properties of the particular machine, which can only be provided by a thorough lattice model. A lattice model consists of all the accelerator elements that affect the beam motion and markers of critical locations at which beam parameters may need to be evaluated. Beam motion in the machine can be predicted with the lattice model by sequentially calculating the effects of the individual elements on the particles. One only needs to improve the accuracy of the modeling of the individual elements in order to accurately describe the beam motion and the beam properties in a complex machine.

The beam motion through an element can be described either with a transfer map, or by directly tracking phase space coordinates of particles. For the transfer map approach to achieve a high accuracy, it is necessary to expand to the higher orders, which makes the map cumbersome and slow to evaluate. Transfer maps can also be used to track particles. However, the tracking results are not symplectic as either the map is not symplectic (e.g., Taylor maps) or the map needs to be truncated during evaluation (Lie maps). Hence tracking with transfer maps is not suitable for the study of long-term stability. In practice, it is more common to use element-by-element particle tracking.

In element-by-element particle tracking, the 6-dimensional phase space coordinates, $(x, p_x, y, p_y, z, \delta)^T$ (or 4-dimensional if the longitudinal motion is not included), are passed through each element from the entrance face to the exit face. The coordinate changes in an element depend on the physical process involved, the element parameters, and the initial coordinates. Using particle tracking, the lattice features and beam properties can be evaluated.

2.7.1 Tracking different types of accelerator elements

Linear elements:

Tracking through the linear elements, namely, drift spaces, dipole magnets, and quadrupole magnets, is straightforward. The transfer matrices in Eqs. (1.27), (1.32), and (1.37) (or the corresponding forms with negative gradients) can be directly applied. In these equations, the transfer matrices are given for x' and y' coordinates, instead of p_x and p_y. Eq. (1.2) can be used to convert between the coordinates for each particle before and after the matrices are applied. The gradients and the curvature in Eqs. (1.32) and (1.37) are scaled with $\frac{1}{1+\delta}$, hence the effects of chromatic errors on the beam motion are included in the tracking.

The only effect on the longitudinal coordinates by these linear elements is a shift in z. This can be derived by integrating Eq. (1.12). The z shift in drift spaces and quadrupoles are second order functions of the transverse coordinates at the entrance. For dipole magnets, the transverse motion in the bending plane (typically horizontal) is coupled to the longitudinal motion. The 6×6 transfer matrix has non-zero R_{16}, R_{26} elements, as given by $\mathbf{d}$ in Eq. (1.33). R_{51} and R_{52} are non-zero and are connected with R_{16} and R_{26} through Eq. (2.47). The R_{56} element is also non-zero. For example, the linear terms of Δz for a pure sector dipole are given by

$$\Delta z = \frac{\delta s}{\gamma_0^2} - x_0 \sin(hs) - \frac{x_0'}{h}(1 - \cos(hs)) + \delta(s - \frac{1}{h}\sin(hs)), \qquad (2.118)$$

where subscript 0 indicates coordinates at the entrance face.

Edge focusing for the dipole magnets can be included by applying the transfer matrix in Eq. (1.35) with the proper entrance and exit angles. To account for the finite extent of the fringe field, the vertical angle may be corrected using the fringe field integral and the bending radius [17]. Effects of fringe fields in quadrupoles may also be included. The dominant effect of the soft-edge gradient variation is to scale up one transverse coordinate and scale down its conjugate coordinate [66].

Symplectic integration:

Sextupole magnets and higher order multipoles are nonlinear elements. There is no closed-form analytic solution to the motion in these elements. Application of ordinary numeric integration to solve the equations of motion through such elements can yield accurate results for one pass. However, the solution does not preserve symplecticity of the particle motion and is thus not ideal for long-term tracking simulation. Symplectic integration [101, 38] has to be used for particle tracking in nonlinear elements.

The general idea of explicit symplectic integration is to split the Hamiltonian into two integrable parts, such as a drift space and a lumped kick. In each integration step, a number of drifts and kicks are alternately applied to the particles. Since transporting through drifts and thin-lens kicks are both symplectic, the total transport is automatically symplectic. If the lengths of the drifts, the strengths of the kicks, and the order of application are properly chosen (independent of the actual Hamiltonian), the integration will be accurate as well as symplectic. A simple case is the second order symplectic integrator illustrated in Figure 2.7. Each integration step consists of a drift, a kick, and another drift. The lengths of the drifts are equal to one half of the step length and the kick corresponds to the integrated magnetic field over the step length. The symmetric configuration eliminates the first order errors such that the leading error terms are $O(L^2)$. Slicing the element into many integration steps will increase the accuracy of the solution.

The commonly used fourth order symplectic integrator [38] is composed of four drifts and three kicks in each step. The lengths of the four drifts (in the order of occurrence) are $\alpha_1 L$, $\alpha_2 L$, $\alpha_2 L$, and $\alpha_1 L$, respectively, with $\alpha_1 = \frac{1}{2(1+\zeta)}$, $\alpha_2 = \frac{\zeta}{2(1+\zeta)}$, and $\zeta = 1 - 2^{1/3}$. The three kicks are inserted between the drifts and their strengths corresponds to integration lengths of $\beta_1 L$, $\beta_2 L$, and $\beta_1 L$, with $\beta_1 = 2\alpha_1$ and $\beta_2 = 2(\alpha_2 - \alpha_1)$. Note $\alpha_2 < 0$ and $\beta_2 < 0$ and hence the corresponding drift and field lengths are negative.

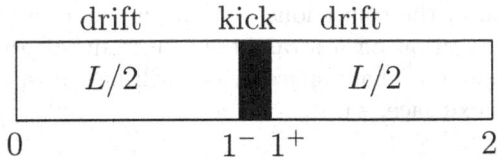

Figure 2.7 A second order integrator that consists of two drifts and a lumped kick in the middle. The kick strength corresponds to the integrated field over the length.

To model higher order multipole components in quadrupole and dipole magnets due to systematic or random errors, symplectic integration is also used. Application of the symplectic integrators to straight elements (i.e., $h = 0$) is straightforward as the drift space terms and the magnetic field terms are naturally separated. The Hamiltonian (with the hard-edge field model) is in the form

$$H = H_1 + H_2 = (1 + \delta) - \sqrt{(1 + \delta)^2 - p_x^2 - p_y^2} - a_s(x, y), \qquad (2.119)$$

where $H_2 = -a_s(x, y)$ represents the magnetic fields and H_1 represents the drift space. Typically the small angle approximation, $H_1 \approx \frac{p_x^2 + p_y^2}{2(1+\delta)}$, can be used.

The separation of the Hamiltonian in a dipole magnet is more involved. The Hamiltonian for a dipole element is given by

$$H = (1 + \delta) - (1 + hx)\sqrt{(1 + \delta)^2 - p_x^2 - p_y^2} - (1 + hx)a_s(x, y), \qquad (2.120)$$

which can also be split into the potential $H_2 = -(1+hx)a_s(x, y)$ and the drift space, with $H_1 = (1 + \delta) - (1 + hx)\sqrt{(1 + \delta)^2 - p_x^2 - p_y^2}$. The solution to the motion in a drift space in the curved reference system is [37],

$$x_2 = (\rho + x)\frac{\cos\phi}{\cos(\phi + hL)} - \rho, \qquad (2.121a)$$

$$p_{x2} = \sqrt{(1 + \delta)^2 - p_y^2}\sin(\phi + hL), \qquad (2.121b)$$

$$y_2 = y + p_y\frac{(\rho + x)}{\sqrt{(1 + \delta)^2 - p_y^2}}(\cos\phi\tan(\phi + hL) - \sin\phi), \qquad (2.121c)$$

$$z_2 = z + (1 + \delta)\frac{(\rho + x)}{\sqrt{(1 + \delta)^2 - p_y^2}}(\cos\phi\tan(\phi + hL) - \sin\phi) - L, \qquad (2.121d)$$

where subscript 2 indicates values at the exit face and

$$\phi = \tan^{-1}\frac{p_x}{\sqrt{(1 + \delta)^2 - p_y^2}}.$$

The approximation of the continuous bending in the curved coordinate system with the propagation on a straight line introduces errors. Notably, an on-energy particle at the phase space origin will be transported to non-zero coordinates at the exit face, with

$$\Delta x_0 = \frac{1}{\cos hL} - \rho, \ \Delta p_{x,0} = \sin hL, \ \Delta z_0 = \rho \tan hL - L. \qquad (2.122)$$

These errors need to be subtracted from Eq. (2.121). There are other ways to separate the dipole Hamiltonian for symplectic integration. Sometimes the simple approach of replacing Eq. (2.121) with the solution of a drift space in the straight coordinate system is used. The evaluation is faster, although it comes with some loss of accuracy.

RF cavities:

An RF cavity can be modeled as a thin-lens element in which the momentum deviation coordinate is modified according to the z-coordinate of the particle and the RF parameters, which gives

$$\delta_2 = \delta + \frac{eV}{\beta^2 E_0} \sin(\frac{2\pi h f_0 z}{c} + \phi_s), \qquad (2.123)$$

where V is the RF voltage, f_0 the revolution frequency, h the harmonic number, and ϕ_s is the synchronous phase.

Radiation damping and quantum excitation:

Radiation damping and quantum excitation occur in electron storage rings in which particles emit photons due to synchrotron radiation [106]. Since higher energy particles lose more energy to photons, lower energy particles lose less, and all particles on average gain the same amount of energy each turn, the energies of all particles tend to converge to the same value. Similarly, as particles lose the transverse momenta to photon emissions and only gain energy through work done in the longitudinal direction, the transverse oscillations gradually decrease toward zero. Radiation damping can be implemented in the dipole symplectic integrator by making the particles lose the correct amount of energy after each kick, and scale p_x and p_y to keep the x' and y' coordinates unchanged. With radiation damping, the beam motion is no longer symplectic.

The emission of each photon gives the particle a kick in the momentum coordinate. The kicks by the emission of photons put the particle on a random walk in the longitudinal phase space, resulting in increasing longitudinal action variable. When photon emissions occur in a dispersive region, the particle will start to oscillate around the off-energy closed-orbit corresponding to the new energy; hence the particle is excited in the transverse plane. The excitation of beam motion by the impulsive kicks of photon emissions is called quantum excitation or quantum diffusion. Quantum excitation causes particle to deviate from the closed-orbit. It is balanced by radiation damping, leading to an equilibrium beam distribution. To simulate quantum excitation, the energy loss of the particle at each kick is given by a random variable, which can be drawn from a Gaussian distribution with the proper mean and standard deviation. Since the use of kicks with negative field lengths introduces extra excitation, the second order symplectic integrator, in which the kick length is positive, is preferred.

Misalignment:

There are always errors in the positions and orientations of the accelerator elements in a real machine as compared to the ideal design. These errors are referred to as misalignment. Misalignment of magnets can significantly impact the accelerator performance. For example, transverse position shifts of quadrupole magnets produce dipole kicks to the beam through the feed-down effects. The kicks cause closed orbit offsets, typically much larger than the alignment errors themselves. The ratio of the induced rms orbit offset to the rms misalignment of magnets is called the amplification factor, which typically ranges from 10 to 100. Small misalignment errors, such as 100 μm, can cause large orbit errors on the order of 1 to 10 mm and in turn optics errors, coupling, and degradation of nonlinear dynamics performance.

Modeling of small alignment errors can be done by performing coordinate transformations at the entrance and exit faces of the misaligned elements. For example, for a horizontal alignment error of Δx, the x-coordinate of the particle is first shifted by $-\Delta x$ at the entrance, and after tracking through the element, shifted back by Δx at the exit. When multiple alignment errors are modeled for one element, the transformations are applied in the opposite order at the exit and entrance faces.

2.7.2 Calculation of lattice functions and beam parameters

With the ability to track phase space coordinates of particles through the lattice, various lattice functions can be calculated.

Closed-orbit:

The closed-orbit can be found by solving for a coordinate vector that satisfies the fixed-point condition $\mathbf{M}(\mathbf{X}_c) = \mathbf{X}_c$, where $\mathbf{M}$ represents the one-turn map, here executed by particle tracking simulation.

There are two scenarios. First, the lattice consists of no RF element and a closed-orbit is found for the on-energy particle or a particle with a given momentum deviation, δ. The 4-dimensional closed-orbit vector, $\mathbf{X}_c = (x, p_x, y, p_y)_c^T$, can be found iteratively. At each iteration, the present solution, $\mathbf{X}_n$, is tracked for one-turn. The solution for the next iteration is then given by

$$\mathbf{X}_{n+1} = \mathbf{X}_n + [\mathbf{I} - \mathbf{R}_4(\mathbf{X}_n)]^{-1}(\mathbf{M}(\mathbf{X}_n) - \mathbf{X}_n), \qquad (2.124)$$

where $\mathbf{R}_4(\mathbf{X}_n)$ is the 4×4 transfer matrix on orbit $\mathbf{X}_n$ (which can be approximated with $\mathbf{R}_4(\mathbf{X}_0)$), and the longitudinal coordinates of $(0, \delta)$ are used in tracking. The initial solution may be set to $\mathbf{X}_0 = (0, 0, 0, 0)^T$.

In the second scenario, the lattice can have RF elements and other elements that change the beam energy, for example, radiation damping or impedance elements. The goal is to find the 6-dimensional closed-orbit. A particle on this closed-orbit is synchronous with the RF cavities. The same iterative procedure as in Eq. (2.124) can be used, except now the one-turn transfer matrix and the orbit vectors are now 6-dimensional.

Linear optics functions:

The transfer matrix between any two locations in the lattice can be computed with numeric differences of the particle coordinates. For example, the i'th column of the transfer matrix (i.e., the linear dependence of the exit coordinates on the i'th coordinate at entrance) is calculated by tracking two particles whose initial i'th coordinate is shifted by $\pm\Delta$, respectively,

$$\mathbf{R}_{:i} = \frac{\mathbf{M}(\mathbf{X}_2) - \mathbf{M}(\mathbf{X}_1)}{2\Delta}, \tag{2.125}$$

where $\mathbf{X}_2$ and $\mathbf{X}_1$ are equal to the reference orbit, $\mathbf{X}_0$, except their i'th components are given by $\mathbf{X}_2(i) = \mathbf{X}_0(i) + \Delta$, $\mathbf{X}_1(i) = \mathbf{X}_0(i) - \Delta$. Numerically $\Delta = 1 \times 10^{-8}$ may be used for double-precision computers.

The one-turn transfer matrix for a ring lattice can be similarly computed. The matrix is usually calculated on the closed-orbit. For uncoupled lattices (or with weak x-y coupling), the 2×2 matrices for the horizontal and vertical planes can be used to calculate the betatron tunes and the Courant-Snyder parameters, using Eq. (1.58). The C-S parameters at other locations and the phase advances can be calculated with Eq. (1.70) and (1.71), respectively, using the transfer matrix between the two locations. Parametrization of coupled motion can be done with the procedure described in Eqs. (2.59-2.63).

The dispersion functions at one location can be calculated from the one-turn transfer matrix using Eq. (2.37). Dispersion functions elsewhere can be obtained by transporting the dispersion vector with the extended transfer matrix. The momentum compaction factor can be calculated from the R_{56} element of the one-turn transfer matrix with Eq. (2.49).

By calculating the transfer matrix with a small fixed momentum deviation and in turn the corresponding betatron tunes, the chromaticities can be obtained. Chromatic beta beating can also be calculated with the off-energy transfer matrix.

Nonlinear beam dynamics performance:

The nonlinear beam dynamics of a circular accelerator can be characterized with particle tracking simulation. Basic nonlinear dynamics features, such as betatron tune shifts with oscillation amplitudes and momentum deviation, are typically computed for fixed-momentum particles (with RF cavities turned off in the lattice). In these calculations, particles with a series of initial x, y, or δ coordinates (while all other 5 coordinates are equal) are launched and tracked for a number of turns (e.g., 1024 turns). For example, for tune shifts with the x-amplitude, particles with initial coordinates $y = 0.1$ mm, $p_x = p_y = z = \delta = 0$, and $x = -10$ mm to 10 mm with a step size of 0.25 mm may be launched for a ring with a dynamic aperture around 10 mm. A small initial y offset is used for the purpose of evaluating the vertical tune from the orbit oscillations. Betatron

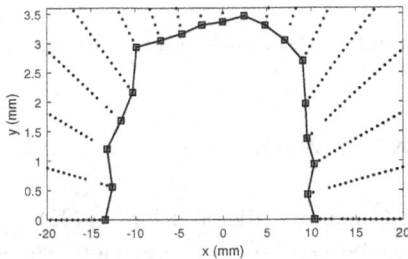 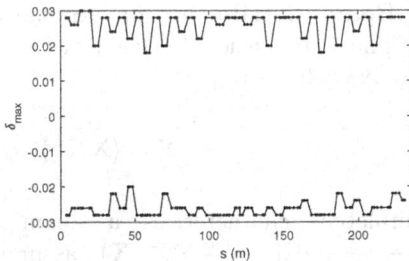

Figure 2.8   The DA (left) and LMA (right) for the SPEAR3 7-nm lattice. The dots in the left plot show the lost particles. The LMA gives the maximum momentum deviation error for a particle launched from a location without being lost.

tunes can be determined from turn-by-turn position data with high precision methods such as NAFF [75] and interpolated FFT [7] (see Chapter 5).

Frequency map analysis (FMA) [74] is often performed to study the non-linear beam dynamics for storage rings. To compute the x-y frequency map, particles are launched from a grid on the x-y plane (typically with $y \geq 0$) that extends to beyond the edge of the dynamic aperture. The betatron tunes are evaluated for each particle. Also evaluated is the betatron tune diffusion rate, defined as $\frac{1}{2N} \log_{10}(\Delta\nu_x^2 + \Delta\nu_y^2)$, where $\Delta\nu_{x,y}$ are tune differences from the first N turns to the second N turns for the two transverse planes, respectively. Plotting the tune diffusion rate in the x-y plane and the tune diagram can reveal the nonlinear resonances that limit the dynamic aperture (DA). The x-δ frequency map can be similarly computed to study the motion of off-energy particles.

Ultimately the nonlinear beam dynamics performance of a storage ring concerns the DA and local momentum apertures (LMA). To evaluate the DA and LMA, the lattice conditions are made as realistic as possible, with RF cavities and radiation damping included. Typically, the dynamic aperture is computed by launching particles equally distributed on a number of rays in the x-y plane extending from the origin and tracking for many turns. For electron storage rings, the number of turns is usually comparable to the damping time. Physical apertures may be included in the lattice to intercept particles with large position offsets. The boundary defined by connecting the last surviving particle before the first lost particle on each ray is the dynamic aperture.

The LMA usually is calculated for a variety of representative locations in a periodic cell. At each location, particles with initial momentum deviations covering the potential aperture boundary are launched and tracked for many turns. The boundaries at both the negative side and the positive side are obtained by connecting the initial δ-coordinate of the last surviving particle before the first lost particle. Figure 2.8 shows the DA and LMA for a SPEAR3 upgrade lattice.

Beam distribution parameters:

The 6-dimensional second order moment matrix (also known as the Σ-matrix) defined by

$$\Sigma = \langle \mathbf{X}\mathbf{X}^T \rangle = \int \mathbf{X}\mathbf{X}^T \rho(\mathbf{X})d\mathbf{X}, \qquad (2.126)$$

completely characterizes a beam in Gaussian distribution, $\rho(\mathbf{X}) = \frac{1}{(2\pi)^3\sqrt{\det \Sigma}} \exp(-\frac{1}{2}\mathbf{X}^T\Sigma^{-1}\mathbf{X})$ (assumed to be centered on the reference orbit for notation simplicity, i.e., with $\langle \mathbf{X} \rangle = 0$). The Σ-matrix is also often used to characterize beams in other distributions. If $\mathbf{M}$ is the transfer matrix between two locations, $\mathbf{X}_2 = \mathbf{M}\mathbf{X}_1$, it is straightforward to show that $\Sigma_2 = \mathbf{M}\Sigma_1\mathbf{M}^T$. For a symplectic matrix $\mathbf{M}$ (see Eq. (1.44)), it follows that $\Sigma_2\mathbf{S} = \mathbf{M}\Sigma_1\mathbf{S}\mathbf{M}^{-1}$ and hence the eigenvalues of $\Sigma\mathbf{S}$ do not change in symplectic transportations [120]. It can be shown that the 6 eigenvalues are $\pm i\epsilon_k$, $k = 1, 2, 3$, and $\det \Sigma = \epsilon_1^2\epsilon_2^2\epsilon_3^2$, where ϵ_k are the eigen-emittances for the three degrees of freedom of particle motion. In the typical case of weak coupling, the eigen-emittances correspond to the three planes, x, y, and z, respectively.

In electron storage rings, the beam reaches an equilibrium distribution due to radiation damping and quantum excitation. For an uncoupled lattice, the emittances can be calculated from radiation integrals, which are determined by the linear lattice functions [106]. However, it becomes more complicated when there exists linear coupling between the horizontal and vertical planes. A general procedure can be applied to track through the lattice elements to obtain the transfer matrix with radiation damping and the accumulated quantum diffusion effects, which are then used to solve for the Σ-matrix [90].

The equilibrium distribution can also be found by long-term multi-particle tracking. Radiation damping, quantum excitation, and RF cavities are turned on in the lattice. A large number of particles (e.g., ≥ 1000) are launched with random initial coordinates and are tracked for a few damping times. The particles will settle down to an equilibrium distribution independent of the initial conditions. The 6-dimensional Σ-matrix can be evaluated by replacing the integral in Eq. (2.126) with a summation over all particles, from which the emittances and beam sizes can be calculated.

II

Beam-based Correction

II

Orbit and trajectory correction

The path of the beam through an accelerator lattice may be called the orbit or the trajectory. In circular accelerators, there exists a closed beam path, which is called the orbit. In one-pass lattices, such as linacs or transport lines, the beam path varies from shot to shot, depending on the launching angles and positions. The beam path in one-pass systems is the trajectory, which is often loosely referred to as the orbit.

The beam orbit or trajectory is important to the beam performance. Most beam applications require precise positioning of the beam in order to be effective. For example, in collider experiments, the two colliding beams have to overlap in space for collisions to happen. In synchrotron light sources, the electron beam orbit has to be precisely controlled for the photon beam to be focused on the sample. In a free electron laser (FEL), trajectory errors in the undulators not only affect the photon beam positioning on the samples, but also can severely reduce the FEL power. Orbit or trajectory errors may cause a reduction of the effective aperture when the beam is driven toward the vacuum chamber and cause beam losses. In circular accelerators, orbit errors in nonlinear magnets (e.g., sextupoles) cause optics errors and linear coupling and can impact the nonlinear beam dynamics performance. Typically, orbit or trajectory control is a high priority in accelerator operation.

Orbit or trajectory correction needs several prerequisites. First, there must be diagnostics to measure the beam orbit. This is usually done with beam position monitors (BPMs). It is possible to use beam screens, scanning wires, cameras, or other diagnostics. Second, there must exist steering magnets that can effectively alter the beam orbit. The steering magnets can be dedicated short dipole magnets, correction windings on the main bending magnets, or the main bending magnets themselves. The steering magnets are the actuators or knobs for orbit and trajectory correction. Third, the target orbit or trajectory must be determined. This could be done empirically. Finally, and very importantly, there need to be effective algorithms to calculate the required changes on the actuators for the orbit to move toward the target.

Orbit correction for circular accelerators in the early days was based on the harmonic analysis of the orbit errors (HARMON) or by iteratively applying the most effective knobs (MICADO) [5]. These methods target the global distribution of orbit errors. Orbit stability for photon beams in synchrotron light sources was initially achieved with local orbit bumps [46]. Real-time global orbit correction in light sources was first implemented with the harmonics correction approach [127, 126]. Singular value decomposition (SVD) [41, 42] was introduced for global orbit correction later [21, 23]. With a large number of precise and fast BPMs and correctors distributed throughout the beam line, orbit stability can be achieved with high precision in modern accelerators using the SVD-based method.

In this chapter we first briefly discuss orbit measurement, steering magnets, and the determination of orbit target. The main focus is the SVD-based method for orbit and trajectory correction.

3.1   ACCELERATOR COMPONENTS FOR ORBIT CORRECTION

3.1.1   Beam position monitors

Beam positions in the vacuum chamber of accelerators are typically measured with BPMs. Position measurement with BPMs is non-invasive to the beam. A BPM uses electromagnetic pick-ups attached to the vacuum chamber to detect the image current of the beam. The strengths of the image current signals depend on the proximity of the beam to the pick-ups. When the pick-ups are arranged symmetrically about chamber center, any deviation of the beam position from the center will be reflected on the signal strengths of the pick-ups. The signal difference normalized by the total signal strength can thus serve as a measurement of the beam position.

The pick-ups for BPMs can be button shaped or strip-lines. Button BPMs are common in electron accelerators. A button BPM may consist of four electrodes diagonally arranged on the vacuum chamber. The buttons are typically rotated by 45° from the horizontal mid-plane in electron storage rings to avoid synchrotron radiation damage. This is illustrated in Figure 3.1 for a round vacuum chamber.

The signal strength on each button when a beam with an offset from the center passes can be calculated by integrating the surface charge density over the extent of the button. The image charge density on a circular vacuum chamber is given by

$$\sigma(\phi) = -\frac{\lambda}{2\pi R} \frac{R^2 - r^2}{R^2 + r^2 - 2rR\cos(\phi - \theta)}, \tag{3.1}$$

where R is the radius of the vacuum chamber, λ is the line density of the beam current, and (r, θ) is the beam center in polar coordinates. The sum signal, defined as the sum of signals on all four buttons,

$$\Sigma = A + B + C + D, \tag{3.2}$$

is proportional to the beam intensity and the sum of the angles subtended by the buttons. The horizontal and vertical difference signals are defined as

$$\Delta_x = A + D - B - C, \qquad \Delta_y = A + B - C - D, \tag{3.3}$$

respectively. The difference signals depend on the position of the beam center and the beam intensity. It can be shown that the ratio of the difference and sum signals are related to the position of the beam center by

$$\frac{\Delta_x}{\Sigma} = \sqrt{2}\frac{\sin\Delta}{R\Delta}x + O(r^3), \qquad \frac{\Delta_y}{\Sigma} = \sqrt{2}\frac{\sin\Delta}{R\Delta}y + O(r^3), \tag{3.4}$$

where Δ is one half of the angle subtended by each button and $O(r^3)$ represent terms of r^3 or higher. Therefore, the beam position can be calculated from the ratios with a linear conversion. This works well in the vicinity of the center of the vacuum chamber. However, when the beam is far from the center, the higher order terms in Eq. (3.4) will become important and hence the beam position reading obtained with the linear conversion will deviate from the actual beam position. The "measured" beam position vs. the actual beam position for a round circular chamber is shown in the right plot of Figure 3.1.

In synchrotron light sources, the cross section of the vacuum chamber in the arcs is typically not circular. In that case the signals on the BPM buttons need to be calculated with the proper boundary conditions. The ratio of the difference and sum signals will still be a good indication of the beam position in the vicinity of the chamber center, although the conversion coefficient may differ. The nonlinear response of the BPM reading to a large beam position offset from the chamber center will also be different. Figure 3.2 shows the configuration of the BPM buttons in the SPEAR3 vacuum chamber (left) and the nonlinear response of the BPMs to the horizontal position offset [112].

The raw signals from the pick-up electrodes consist of a series of pulses that correspond to the beam bunch passes. When processed through fast electronics, these signals could potentially yield the beam position of each bunch.

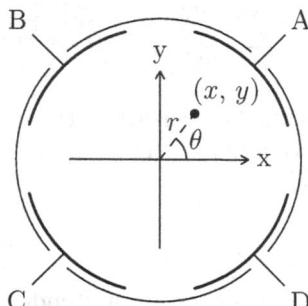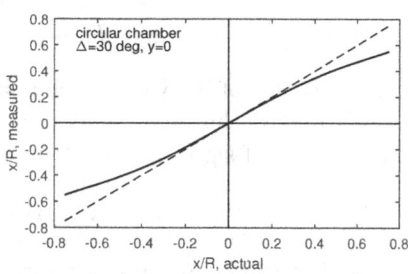

Figure 3.1 Left: BPM button configuration in a round vacuum chamber; Right: the nonlinear response of the BPM readings for the round chamber configuration with a half button width of $30°$.

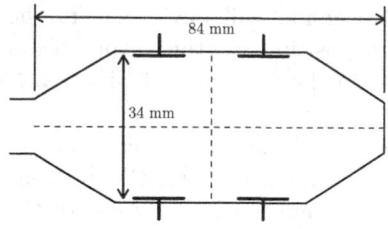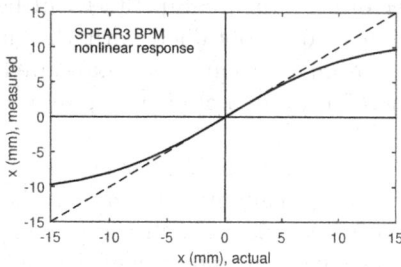

Figure 3.2 Left: BPM button configuration in the SPEAR3 vacuum chamber; Right: the horizontal nonlinear response of the SPEAR3 BPMs.

In linacs or transport lines, the BPMs report beam positions on each beam pass. In storage rings, the repetition rate of bunches is typically on the order of a few hundred MHz. Modern BPM electronics on most of the new rings can resolve the beam position on turn-by-turn basis. The turn-by-turn BPM readings report the average position of all bunches in the ring on the same pass. Turn-by-turn BPMs can monitor the beam motion of bunches moving in phase, such as in the typical case of betatron oscillations when the bunches in the beam are launched with common position and angle coordinates, or when the beam is kicked by a pinger. The turn-by-turn beam positions measured by the BPMs for a beam undergoing betatron oscillations contains information of the linear optics of the machine.

The BPMs also report the average beam position over many turns. This is the typical mode of operation for BPMs during beam delivery to users. The average positions can be reported at a high frequency (e.g., 10 kHz), which can be used for fast orbit correction. When the beam is not in an excited state (synchrotron motion, betatron motion, or instability), the orbit is typically dominated by low-frequency motion. Because the beam in betatron motion oscillates around the closed orbit in high frequency, the time averaged position is a good measurement of the closed orbit as the oscillation is cancelled out. The average beam position have a high precision. It is common for the position readings to have a standard deviation (noise sigma) at or below 1 μm.

In an accelerator, BPMs are distributed throughout the beam path. BPMs are required at certain critical locations in the machine, for example, near the interaction points in a collider, or at the ends of straight sections that host undulators in synchrotron light sources. In other areas, BPMs are placed to sufficiently sample the orbit errors in order to prevent undetected orbit drifts. Orbit responses or betatron oscillations measured by BPMs can also be used for linear optics calibration. It is advisable to have a few BPMs in each betatron period. In a periodic lattice, the BPMs are usually located at identical positions in each cell.

3.1.2 Orbit correctors

Orbit correctors are used to steer the beam orbit toward the desired target orbit or trajectory. The steering is achieved by generating or changing a dipole field on the path of the beam. The dipole field deflects the direction of the propagation of the beam. The deflection is called a kick. The trajectory downstream of the kick is thus modified. In a circular accelerator, a kick changes the closed orbit all around the ring.

An orbit corrector can be a standalone magnet, a set of wires on a multipurpose magnet, or trim coils on a main dipole magnet. Orbit correction can also be achieved by modifying the setpoint of the main dipoles. Trim coils or main dipoles have slow responses to changes due to the large magnetic inductance of the main dipole magnets. It could take seconds for the magnets to settle on the new setpoints. Therefore they cannot be used to correct fast orbit disturbances. In machines that require high orbit stability, such as synchrotron light sources, orbit correctors with fast responses are needed.

Orbit errors are caused by deviations of the bending fields along the beam path from the design condition. The bending field deviations are the sources of orbit errors. Ideally, the error sources should be corrected locally at the source points, such that no orbit distortion occurs elsewhere. However, in reality, local correction is often not practical. This is because the error sources cannot be identified or no adjustment to the bending field can be made to compensate the errors at the locations. In such cases, global orbit correction through a distribution of orbit correctors is desired.

The differential orbit shift caused by a small change to the corrector strength is the orbit response of the orbit corrector. Orbit responses at the BPMs can be measured by changing the corrector current by a small amount and monitoring the orbit shifts. The orbit responses at multiple BPMs of multiple orbit correctors can be arranged in a matrix, called the orbit response matrix, whose element is given by

$$R_{ij} = \frac{\Delta x_i}{\Delta \theta_j}, \tag{3.5}$$

where Δx_i is the orbit change at BPM i for a kick angle, $\Delta \theta_j$, at corrector j. With M BPMs and N correctors, the dimension of the orbit response matrix is $M \times N$. The orbit response matrix can be used to predict the orbit changes when the orbit correctors are varied. Conversely, knowing the desired orbit changes, the orbit response matrix can be used to calculate the required changes to the orbit corrector strengths.

3.2   BEAM-BASED ALIGNMENT

For orbit correction, a target orbit given in terms of BPM readings is needed. The design orbit would be the natural choice for the target orbit. However, it is not straightforward to determine the design orbit on the BPMs. The vacuum chambers, to which the BPMs are attached, have alignment errors. The BPM buttons have mechanical errors in the sizes and positions. The BPM electronics also have errors in signal processing. Therefore, a BPM reading of zero does not represent the design orbit, nor does it represent the mechanical center of the BPM.

On the other hand, since all the magnets have alignment errors from their design positions, the design orbit is not necessarily the ideal orbit for the beam for the purpose of optimizing beam performance. At certain locations of the beam line, the beam position affects the experiments that use the beam. For example, the position and angle of the electron beam in an insertion device affect the photon beam position. For such locations, the target particle beam orbit on the BPMs would be determined by the requirements of the user experiments. In other areas of the beam path, however, there is some freedom in choosing the target orbit. In general, the target orbit should be chosen for the beam to go through the good field regions of the magnets and to maintain a large acceptance by the physical apertures.

A good choice of the target orbit is to steer the beam through the magnetic centers of the quadrupole magnets. The magnetic center of a quadrupole is where the magnetic field crosses zero. It is typically the center of the good field region in the magnet. When the beam goes through the quadrupole center, it receives no angular kick from the magnet and hence the beam orbit is not altered, regardless of the strength of the quadrupole magnet. One benefit of such an orbit is that the strength of the quadrupole magnet, and hence the linear optics, can be changed, without perturbing the beam orbit. In other words, the control of linear optics is decoupled from the control of the beam orbit.

The magnetic center of a quadrupole can be found through a procedure called beam based alignment (BBA) [94]. BBA is based on the very fact that the beam receives no kick at the quadrupole center. In the procedure the BPM nearest to the quadrupole is used to register the quadrupole center. An orbit corrector in the machine is used to change the orbit at the quadrupole. The betatron phase advance between the corrector and the quadrupole needs to be at an appropriate value for the corrector to be effective. At each orbit, the strength of the quadrupole is changed to $I_0 - \Delta I$ and then $I_0 + \Delta I$, while the orbit shifts due to the quadrupole strength step change are recorded by all available BPMs. For each observing BPM j, the orbit shifts induced by the quadrupole modulation, Δx_{ij}, can be plotted against the orbit reading on the nearest BPM, x_i, for the i'th corrector induced orbit. The data points will form one line for each BPM if the orbits are not far from the quadrupole center. With a selection of observing BPMs, there will a collection of lines in the plot, all of which cross at one particular point. The cross point indicates the BPM reading that corresponds to the quadrupole center. Such a plot is called a "bow tie" plot, an example of which is shown in Figure 3.3.

The BBA procedure described in the above does not need a calibration for the quadrupole magnet or the BPMs. Nor does it utilize a lattice model. It finds the quadrupole center directly as measured by the raw reading on the nearest BPM. This procedure works in a circular accelerator as well as in a one-pass lattice. In the latter case, a corrector upstream of the quadrupole is used to change the orbit at the quadrupole, and BPMs downstream are used to detect the orbit shifts due to the quadrupole modulation. The horizontal and vertical offsets are determined separately.

Sometimes there is no corrector magnet conveniently located to alter the orbit at the quadrupole of interest, or sometimes there is no BPM located next to the quadrupole. This would more likely happen in a transport line. In such a situation, it is still possible to find the quadrupole center by stepping its strength and monitoring the orbit changes at downstream BPMs. Using a lattice model between the quadrupole and the BPMs, the kick angle

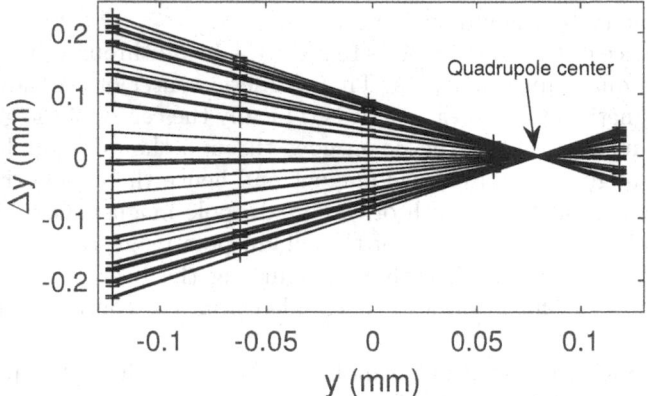

Figure 3.3 Example of a bow-tie plot in beam-based alignment. The horizontal axis is the orbit reading on the nearest BPM to the quadrupole. The vertical axis is the orbit change at other BPMs corresponding to a step change of the quadrupole strength. Each line represents data for one BPM.

corresponding to a step change of the quadrupole current can be derived with

$$\Delta\theta = \frac{\Delta x_i}{R_{iq}}, \tag{3.6}$$

where Δx_i is the orbit shift at BPM i and R_{iq} is the orbit response for a kick at the quadrupole location to the BPM. BPMs close to the quadrupole yet with a sufficiently long lever arm are preferred to avoid effects of lattice errors. Multiple BPMs can be used to derive the kick angle $\Delta\theta$ for a quadrupole current change ΔI. If the current to gradient conversion rate, $r = \frac{\Delta K}{\Delta I}$, and the effective length of the quadrupole, L_q, are known, the orbit offset, x_q, at the quadrupole can be calculated from

$$\frac{\Delta\theta}{\Delta I} = rL_qx_q. \tag{3.7}$$

This approach can also be applied to storage rings, although in this case the accuracy may be impacted by the linear optics errors introduced by the quadrupole strength modulation.

3.3   ORBIT CORRECTION

The goal of orbit correction is to steer the beam orbit with correctors toward the target orbit. The target orbit may be specified only at one or a few selected locations, or throughout the beam path. The former case is referred to as local orbit correction and the latter global orbit correction.

In the earlier days of accelerators, the ability to control beam orbit was limited for reasons such as the lack of orbit monitors and corrector magnets and the lack of effective orbit correction schemes. Global orbit correction could only reduce the overall orbit distortion to a certain level. Local orbit correction was often used to steer the beam at important locations such as the interaction region for a collider, or the source points of light sources. Presently accelerators are equipped with many high precision BPMs and fast and well regulated corrector magnets. The calculation of corrector strengths from the measured orbit errors in global orbit correction with the singular value decomposition (SVD) has been very successful. The beam orbit can often be precisely controlled on target by a global orbit feedback system. Creating a local orbit bump can be as easy as specifying the desired orbit and leaving the rest to the orbit feedback. Local orbit bump can also be easily created using the orbit response matrix with the SVD method. Therefore, we will be focused on the SVD-based global orbit correction method in the following. Early day global orbit correction methods and local orbit correction are discussed briefly toward the end of the section.

There are typically multiple BPMs and multiple correctors, e.g., with M BPMs and N correctors. All three situations, $M > N$, $M = N$, or $M < N$, are possible. When there are more BPMs than correctors (i.e., $M > N$), the system is over-constrained; the orbit target may not be met on all BPMs. As will be discussed later, because of the potential degeneracy in the system, even when $M \leq N$, it is possible that the orbit errors will not be completely eliminated. Therefore, in general, the objective of orbit correction is to minimize the difference between the measured beam orbit and the orbit target in the least square sense, i.e., to minimize

$$\chi^2 = \sum_{i=1}^{M}(x_i - \hat{x}_i)^2, \tag{3.8}$$

where x_i and $\hat{x}_i$ are the measured orbit and the target orbit on BPM i, respectively. Defining the corresponding column vectors $\mathbf{x}$ and $\hat{\mathbf{x}}$ and the residual vector,

$$\mathbf{r} = \mathbf{x} - \hat{\mathbf{x}}, \tag{3.9}$$

the objective function can be expressed as

$$\chi^2 = \mathbf{r}^T \mathbf{r}. \tag{3.10}$$

When the orbit is measured to be $\mathbf{x}_0$, we need to find the desired changes to the corrector magnets, $\boldsymbol{\theta}$, to minimize the objective function, χ^2. Here $\boldsymbol{\theta}$ is a column vector with N elements, whose j'th element, θ_j, represents the desired change of kick angle on corrector j. After the correction $\boldsymbol{\theta}$ is applied to the machine, the orbit will change. Using the orbit response matrix, the new orbit is predicted to be

$$\mathbf{x} = \mathbf{x}_0 + \mathbf{R}\boldsymbol{\theta}, \tag{3.11}$$

and hence the residual vector will become

$$\mathbf{r} = \mathbf{x}_0 + \mathbf{R}\boldsymbol{\theta} - \hat{\mathbf{x}} = \mathbf{R}\boldsymbol{\theta} - \Delta\mathbf{x}, \tag{3.12}$$

where $\Delta\mathbf{x} \equiv \hat{\mathbf{x}} - \mathbf{x}_0$ is the difference between the target orbit and the measured orbit. The objective of the least-square problem, χ^2, is a function of $\boldsymbol{\theta}$,

$$f(\boldsymbol{\theta}) = \chi^2 = (\boldsymbol{\theta}^T \mathbf{R}^T - \Delta\mathbf{x}^T)(\mathbf{R}\boldsymbol{\theta} - \Delta\mathbf{x}). \tag{3.13}$$

The condition for $f(\boldsymbol{\theta})$ to be at a minimum is for its derivatives with all variables θ_j to be zero, i.e.,

$$\frac{\partial f}{\partial \theta_j} = 2\sum_{k=1}^{N} (\mathbf{R}^T \mathbf{R})_{jk} \theta_k - 2\sum_{i=1}^{M} R_{ij} \Delta x_i = 0, \tag{3.14}$$

for $j = 1, 2, \cdots, N$. This can be written in the vector form

$$\frac{\partial f}{\partial \boldsymbol{\theta}} = 2\mathbf{R}^T \mathbf{R}\boldsymbol{\theta} - 2\mathbf{R}^T \Delta\mathbf{x} = 0. \tag{3.15}$$

The solution for $\boldsymbol{\theta}$ is thus

$$\boldsymbol{\theta} = (\mathbf{R}^T \mathbf{R})^{-1} \mathbf{R}^T \Delta\mathbf{x}. \tag{3.16}$$

3.3.1   Orbit correction with SVD

Eq. (3.16) can be used to calculate the desired changes to the corrector magnets for orbit correction when the orbit errors and the orbit response matrix are known. The matrix $\mathbf{R}^T \mathbf{R}$ is $N \times N$ in dimension. Eq. (3.16) gives a unique solution for $\boldsymbol{\theta}$ if and only if the inverse matrix of $\mathbf{R}^T \mathbf{R}$ exists. In reality, this might not be the case. For example, when the number of correctors is larger than the number of BPMs, matrix $\mathbf{R}^T \mathbf{R}$ has a rank $M < N$ and does not have an inverse matrix. Even if mathematically the inverse matrix exists, the solution by Eq. (3.16) might not always be appropriate for orbit correction since matrix $\mathbf{R}^T \mathbf{R}$ could be nearly degenerate, i.e., some of its eigenvalues are very close to zero. In such a case, the solution would be very sensitive to noise in the orbit measurement and can result in large corrector changes in response to small orbit errors. Therefore, a method to solve the least-square problem in Eq. (3.13) with considerations of the realistic conditions is needed.

Singular value decomposition (SVD) [42] of the orbit response matrix provides the answer to the above challenge [21, 23]. The method based on SVD to solve the least-square problem is very powerful and is widely applied to orbit corrections in accelerators. The SVD of the orbit response matrix is in the form of

$$\mathbf{R} = \mathbf{U}\mathbf{S}\mathbf{V}^T, \tag{3.17}$$

with orthogonal matricies $\mathbf{U}$ and $\mathbf{V}$ and diagonal matrix $\mathbf{S}$. With the dimension of $\mathbf{R}$ being $M \times N$, the dimensions of $\mathbf{U}$ and $\mathbf{V}$ are $M \times M$ and $N \times N$, respectively. The dimension of matrix $\mathbf{S}$ is $M \times N$ and its only non-zero elements are $S_{ii} = s_i$, $i = 1, 2, \cdots, \min(M, N)$, where $\min(\cdot)$ represents the smaller of the two numbers. The values s_i are called singular values (SV). The singular values are real numbers greater than or equal to zero. By convention the singular values are ordered in a descending sequence such that

$$s_1 \geq s_2 \geq s_3 \geq \cdots \geq s_{\min(M,N)} \geq 0. \tag{3.18}$$

With $\mathbf{R}$ given in Eq. (3.17), we have

$$\mathbf{R}^T \mathbf{R} = \mathbf{V}(\mathbf{S}^T \mathbf{S})\mathbf{V}^T, \tag{3.19}$$

with

$$(\mathbf{S}^T \mathbf{S}) = \operatorname{diag}(s_1^2, s_2^2, \cdots, s_{\min(M,N)}^2, 0, \cdots, 0), \tag{3.20}$$

where $\operatorname{diag}(\cdot)$ defines a diagonal matrix using the given parameters as its diagonal elements. Here 0's are patched to the end to make the diagonal matrix $N \times N$ in dimension if $M < N$.

In the case $N \leq M$ and all of the singular values of $\mathbf{R}$ are non-zero, then the inverse matrix of $\mathbf{R}^T \mathbf{R}$ exists and is given by

$$(\mathbf{R}^T \mathbf{R})^{-1} = \mathbf{V}(\mathbf{S}^T \mathbf{S})^{-1}\mathbf{V}^T, \tag{3.21}$$

where we used the fact $\mathbf{V}^T = \mathbf{V}^{-1}$ for orthogonal matrix $\mathbf{V}$, and

$$(\mathbf{S}^T \mathbf{S})^{-1} = \operatorname{diag}(s_1^{-2}, s_2^{-2}, \cdots, s_N^{-2}). \tag{3.22}$$

The solution in Eq. (3.16) can now be written

$$\boldsymbol{\theta} = \mathbf{V}\mathbf{S}^{-1}\mathbf{U}^T \Delta \boldsymbol{x}, \tag{3.23}$$

where by definition $\mathbf{S}^{-1}$ is an $N \times M$ matrix given by

$$\mathbf{S}^{-1} \equiv (\mathbf{S}^T \mathbf{S})^{-1}\mathbf{S}, \tag{3.24}$$

whose only non-zero elements are on the diagonal and are simply s_i^{-1}, $i = 1$, $2, \cdots, N$. In this case, the SVD of $\mathbf{R}$ leads to an explicit form of the unique solution to the desired corrector changes, as given in Eq. (3.23).

When $N > M$, or if at least one of the singular values of $\mathbf{R}$ is zero, Eq. (3.22) does not hold since there would be $1/0$'s on the diagonal. In this case, the inverse matrix of $\mathbf{R}^T \mathbf{R}$ does not exist and hence Eq. (3.16) cannot be used. Correspondingly, there is no unique solution to the least-square problem. However, the SVD approach can be used to find an appropriate solution. In Eq. (3.22), we only need to replace the diagonal elements that would be $1/0$'s with zeros, i.e., defining the $N \times N$ pseudo-inverse matrix of $\mathbf{S}^T \mathbf{S}$,

$$(\mathbf{S}^T \mathbf{S})^{-1} = \operatorname{diag}(s_1^{-2}, s_2^{-2}, \cdots, s_{\min(M,N)}^{-2}, 0, \cdots, 0), \tag{3.25}$$

and replacing s_i^{-2} with zeros in the above for any $s_i = 0$. Then Eqs. (3.23)-(3.24) can be used to calculate the solution, $\boldsymbol{\theta}$, in the general case.

The solution Eq. (3.23) can be better understood if we examine the SVD of matrix $\mathbf{R}$ closely. The SVD of $\mathbf{R}$ can be rewritten in the expanded form

$$\mathbf{R} = \sum_{i=1}^{\min(M,N)} s_i \mathbf{u}_i \mathbf{v}_i^T, \qquad (3.26)$$

where $\mathbf{u}_i$ and $\mathbf{v}_i$ are the i'th column in matrix $\mathbf{U}$ and $\mathbf{V}$, respectively. Each term in the summation corresponds to an SVD mode, which consists of the singular value, s_i, and the $\mathbf{u}_i$ and $\mathbf{v}_i$ vectors. The $\mathbf{u}_i$ vector represents the pattern over all BPMs and the $\mathbf{v}_i$ vector represents the pattern over all correctors. The calculation of the predicted orbit shift by a given corrector variation can be written as

$$\Delta \mathbf{x} = \mathbf{R}\boldsymbol{\theta} = \sum_{i=1}^{\min(M,N)} s_i \mathbf{u}_i (\mathbf{v}_i^T \boldsymbol{\theta}), \qquad (3.27)$$

where $(\mathbf{v}_i^T \boldsymbol{\theta})$ is the dot product of the two vectors $\mathbf{v}_i$ and $\boldsymbol{\theta}$. This dot product is a scalar value that represents the projection of the corrector changes to the i'th SV mode. Eq. (3.27) indicates that the projection $\mathbf{v}_i^T \boldsymbol{\theta}$, multiplied by the vector $s_i \mathbf{u}_i$, gives the orbit changes due to the i'th SV mode. Since all the $\mathbf{u}_i$ and $\mathbf{v}_i$ vectors are normalized to $||\mathbf{u}_i|| = ||\mathbf{v}_i|| = 1$, where $|| \cdot ||$ represents the Euclid norm of a vector, the amount of orbit changes with a given projection to the SV mode is determined by the singular value, s_i. A large singular value means the corrector pattern corresponding to the $\mathbf{v}_i$ vector is very effective in changing the orbit (yet only producing changes to BPMs by the pattern as given by the corresponding $\mathbf{u}$ vector). A small singular value means the SV mode is not effective in making orbit changes. In the extreme case, when the singular value is zero, the projection of corrector changes in that mode does not cause any orbit change on the BPMs at all. For a small or zero singular value, the effects of corrector changes with the particular pattern tend to cancel, resulting in small or no net orbit shifts on the BPMs.

The SVD of the orbit response matrix reveals that with a given set of orbit correctors, it is easy to make orbit changes in some patterns, while it is difficult or impossible to make orbit changes in some other patterns. The orbit response matrix can be seen as a map from the N-dimensional corrector space to the M-dimensional BPM space. A vector in the corrector space, $\boldsymbol{\theta}$, is mapped to a vector in the BPM space, $\Delta \mathbf{x}$. The column vectors in the $\mathbf{V}$ matrix represent an orthogonal basis of the corrector space, while the column vectors in the $\mathbf{U}$ matrix represent an orthogonal basis of the BPM space. Corrector changes along the basis vector $\mathbf{v}_i$ result in an orbit shift only along the $\mathbf{u}_i$ vector. The corresponding singular value, s_i, represents the effectiveness of making orbit changes in this mode. Conversely, an orbit shift in the $\mathbf{u}_i$ pattern can only be made through corrector changes in the $\mathbf{v}_i$ pattern (since corrector changes in any other pattern result in an orbit shift that is orthogonal to $\mathbf{u}_i$). Therefore, if we want to make an orbit shift $\Delta\mathbf{x}$, we can first calculate the decomposition of $\Delta\mathbf{x}$ over the BPM modes, and then use the projection on each mode to calculate the required corrector changes. In fact, this is exactly what Eq. (3.23) stands for, which can be re-written as

$$\theta = \sum_{i=1}^{\min(M,N)} \frac{1}{s_i} \mathbf{v}_i (\mathbf{u}_i^T \Delta\mathbf{x}), \tag{3.28}$$

where $(\mathbf{u}_i^T \Delta\mathbf{x})$ is the dot product between $\mathbf{u}_i$ and $\Delta\mathbf{x}$ that represents the projection of the desired orbit shift over the i'th SV mode.

In the case of $N > M$, SVD of the orbit response matrix finds only M basis vectors for the N-dimensional corrector space. There is an extra $N - M$ dimensional subspace of the corrector space that has no impact to the BPM space. In other words, any corrector changes in this subspace do not result in an orbit shift observed on the BPMs. The basis vectors of this subspace are similar to the basis vectors that have zero singular values in terms of not being able to make orbit changes observable on the BPMs. Because corrector changes in the extra subspace do not change the orbit, there is no point of making such corrector changes. This is why we patch zeros to the end in Eq. (3.25). For the same reason, we do not allow corrector changes along the $\mathbf{v}_i$ vectors corresponding to zero singular values. Hence we replace $\frac{1}{s_i}$ with 0 when $s_i = 0$ in Eq. (3.25) and Eq. (3.28).

While in both cases corrector changes cannot cause an orbit shift on the BPMs, there is a subtle difference between the case with a zero singular value and the case with corrector changes in the $N - M$ dimensional subspace. In the latter case, there is a redundancy in the correctors for the set of BPMs. In other words, there are too many correctors and thus there is no unique solution to the orbit correction least-square problem. However, in the former case, with a zero singular value, there is a true deficiency in the orbit correction system such that we lack the ability to make orbit shifts in a certain pattern (i.e., the corresponding $\mathbf{u}_i$ vector).

Similarly, when a singular value is very small, our ability to make orbit shifts in the corresponding BPM pattern is limited. For a small orbit shift in the pattern, large corrector changes are needed, as is evident in Eq. (3.28) by the $\frac{1}{s_i}$ coefficient. Sometimes a singular value is so small, the required corrector changes for an orbit correction could exhaust the strengths of some correctors and cause the orbit correction system to fail. To prevent such a scenario, it is necessary to limit the corrector changes in the SV modes with small singular values. This could be done by choosing a threshold for the singular values and setting $\frac{1}{s_i}$ to zero in Eq. (3.28) for all singular values s_i below the threshold. In doing this we give up the attempt to make orbit shifts in certain patterns in exchange for the stability of the orbit correction system. It is not so much a loss because our ability to make such orbit shifts is limited in the first place.

Figure 3.4 shows the singular value spectra of the orbit response matrices of the horizontal and vertical planes for the SPEAR3 (top plot) and NSLS-II (bottom plot) storage rings. SPEAR3 has 58 horizontal correctors, 56 vertical correctors, and 57 BPMS, distributed over 18 cells. NSLS-II has 180 horizontal correctors, 180 vertical correctors, and 180 BPMs over 30 cells. The betatron tunes are [14.106, 6.177] for SPEAR3 and [33.22, 16.26] for NSLS-II. Each SV spectrum has a pair of leading modes with comparable SV values. The u- and v-vectors of these leading modes resemble betatron orbits of free oscillating particles. Figure 3.5 shows the first SV mode for the SPEAR3 orbit response matrix for both transverse planes. In the leading modes, the corrector strengths vary according to the betatron phase to result in additive contributions to the orbit. The two leading modes have similar spatial patterns but are out of phase by 90°. These two modes contribute strongly to the real and imaginary components of the integer stopband nearest to the betatron tune. At the lower end of the SV spectrum, there is a floor of small singular values. The ratio of the leading SV to the SVs on the floor is about 500 for SPEAR3. The same ratio for NSLS-II is about 2000. The SPEAR3 horizontal orbit response matrix has one extremely small singular value, despite having one more corrector than the BPMs in the plane. This SV mode represents a singularity of the orbit correction system. It corresponds to a local pattern in the double-waist chicane area of the lattice. The singular mode needs to be removed from the calculation of the pseudo-inverse matrix in Eq. (3.25).

An example of storage ring orbit correction with SVD is shown in Figure 3.6. In this example orbit errors were generated in the SPEAR3 lattice by introducing random misalignment errors to all quadrupole magnets, with the rms horizontal offset of 100 μm. The horizontal orbit errors are shown in the top plot. Orbit correction was done with 6 SVs or 56 SVs in the calculation of the pseudo-inverse matrix. The resulting orbits after the corrections are applied to the lattice model are shown in the middle plot. The bottom plot shows the kick angles on all 58 correctors. Using 6 SVs, the rms kick angle of all correctors is 0.29 mrad, which brings the rms orbit distortion from 2.35 mm to 0.71 mm. Using 56 SVs, the rms kick angle is 5 times stronger (at 1.62 mrad), only to bring the rms orbit further down to 0.19 mm. The example shows that the bulk of the orbit errors can be suppressed with only a few SV modes; yet many SV modes are required in order to correct the fine details of the orbit errors.

It is worth pointing out that the predicted residual orbit errors, $x_0 + \mathbf{R}\Delta\boldsymbol{\theta}$, have rms values of 0.72 mm for the 6-SV case and 0.004 mm for the 56-SV case, respectively. The prediction was accurate for the few SV case, but not as good for the 56-SV case. This is because the orbit response matrix depends on the linear optics, which varies with the closed orbit due to the feed-down effect of sextupoles. The orbit response matrix measured around the reference orbit is different from the one around a closed orbit with large distortions. In this case, typically several iterations are applied to correct the orbit. The orbit response matrix does not need to be updated for each iteration. As long

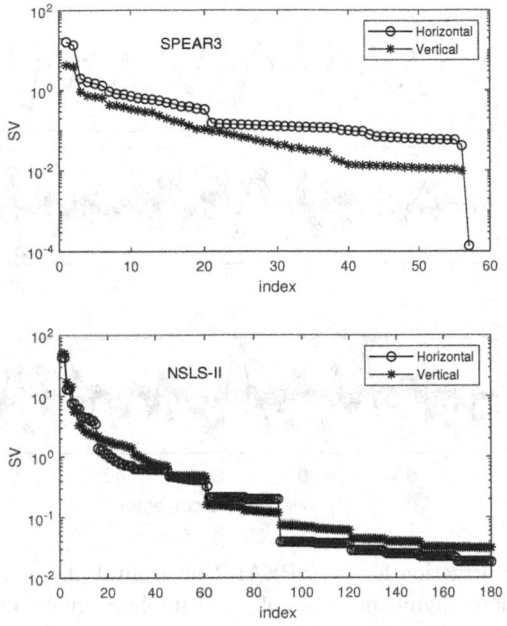

Figure 3.4 Singular values for the orbit response matrices for SPEAR3 (top) and NSLS-II (bottom).

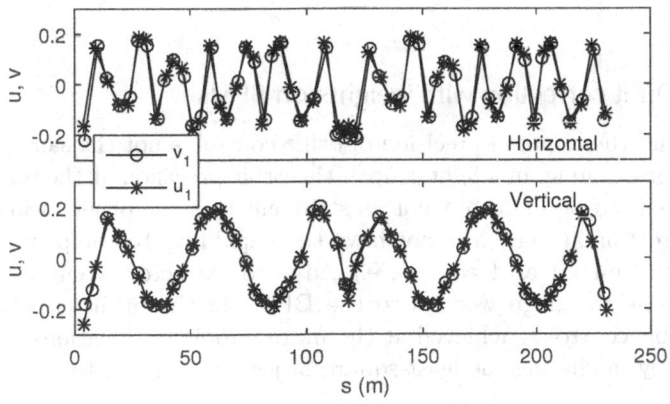

Figure 3.5 The u- and v-vector for the leading SV mode of the SPEAR3 orbit response matrix. Top: horizontal; bottom: vertical.

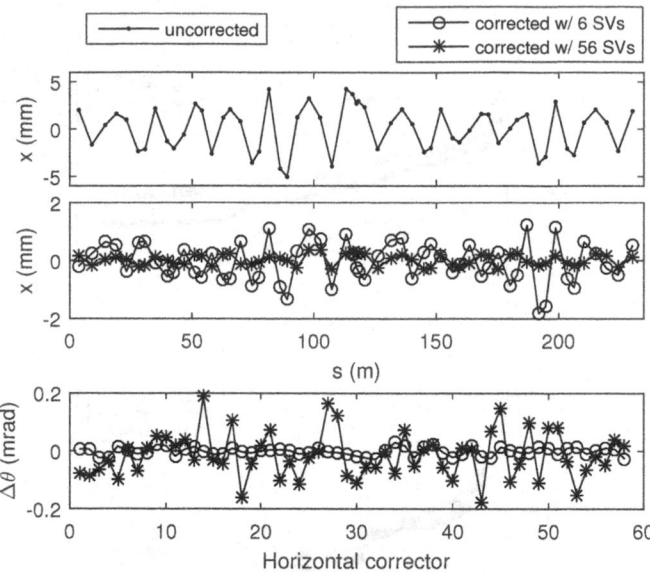

Figure 3.6 Orbit correction for the SPEAR3 horizontal plane. The orbit errors are generated by random alignment errors of quadrupole magnets. Orbit correction was done with the horizontal orbit response matrix, using 6 SVs, or 56 SVs.

as the calculated kick angles are not substantially altered by the errors in the small SVs, in each iteration the orbit distortion will be reduced and hence the relative accuracy of the orbit response matrix will improve, resulting in a converging sequence. To prevent large kick angles due to inaccurate small SVs, sometimes it is necessary to restrict the number of SVs for the first few iterations.

3.3.2 Orbit correction with weights on BPMs

Often times the required precision of orbit control is not the same at different BPMs. For example, in a light source, the orbit precision at the photon source points needs to be high, but not as stringent at some other locations. If the orbit correction system does not have the capability to completely eliminate orbit distortions at all locations, e.g., due to corrector strength limitations, it is sensible to assign weights to the BPMs in the orbit target, such that better orbit control is achieved at the more important locations. This can be achieved by modifying the least-square objective function to

$$\chi^2 = \sum_{i=1}^{M} w_i^2 (x_i - \hat{x}_x)^2, \tag{3.29}$$

where w_i is the weight for BPM i. If the BPMs have different precision, the weight of each BPM could be given as $w_i = \frac{1}{\sigma_i}$, where σ_i is the noise sigma of

BPM i. Additional weight factor could be applied to emphasize the importance of the orbit at certain locations.

In terms of the residual vector, the objective function can now be written

$$f(\boldsymbol{\theta}) = \mathbf{r}^T \mathbf{W}^T \mathbf{W} \mathbf{r}, \tag{3.30}$$

with the diagonal matrix

$$\mathbf{W} = \mathrm{diag}(w_1, w_2, \cdots, w_M). \tag{3.31}$$

Following the same process leading up to Eq. (3.16), the desired corrector change is found to be

$$\boldsymbol{\theta} = (\mathbf{R}^T \mathbf{W}^T \mathbf{W} \mathbf{R})^{-1} \mathbf{R}^T \mathbf{W}^T \mathbf{W} \Delta \mathbf{x}. \tag{3.32}$$

If we define the weighted orbit response matrix,

$$\mathbf{R}_w = \mathbf{W} \mathbf{R}, \tag{3.33}$$

the solution can be rewritten as

$$\boldsymbol{\theta} = (\mathbf{R}_w^T \mathbf{R}_w)^{-1} \mathbf{R}_w^T \mathbf{W} \Delta \mathbf{x}. \tag{3.34}$$

Using $\mathbf{R}_w$ and $\mathbf{W} \Delta \mathbf{x}$ in place of $\mathbf{R}$ and $\Delta \mathbf{x}$, respectively, Eqs. (3.17)-(3.28) can be used for the weighted BPM case.

In the case of $N \leq M$ where all singular values of $\mathbf{R}_w$ are greater than zero and are used in the calculation of the pseudo inverse matrix, the solution to the correctors, $\boldsymbol{\theta}$, is not changed by the weighting factors. However, if $N > M$, or if not all singular values of $\mathbf{R}_w$ are used, then the weighting factors can change the solution to put more emphasis on the BPMs with higher weights.

One application of weighting BPMs in orbit correction is to make large local bumps. To make an orbit bump on a single BPM, the target orbit, $\Delta \mathbf{x}$, is set to zero on all BPMs except the target BPM, which is set to the desired value. Eq. (3.34) can then be used to calculate the corrector changes. If the adjacent BPMs are very close to the target BPM, it could be difficult to make an exact local bump. There are times when we only want to make a large bump at the target, and are not very concerned of the small orbit distortion on the nearby BPMs. In such cases, we can reduce the weights on the adjacent BPMs. Figure 3.7 shows an example, in which we want to make a 1 mm horizontal orbit bump on one of the BPMs in SPEAR3. If all BPMs have the same weights, the solution found with 50 SVs requires kick angles as large as 0.25 mrad. Because the maximum kick angle for one corrector is 1.5 mrad, the maximum bump would be 6 mm. However, if we set the weights of the four nearby BPMs (two on each side) to 20% of the other BPMs, the maximum required kick angle is only 0.125 mrad; the bump on the target BPM can now reach 12 mm.

It is worth pointing out that in the above example, if all 56 SVs are used in the calculation, there will be no difference between the solutions of the no-weight and with-weight cases. It can make a difference if the weights of the nearby BPMs are set to zero, though.

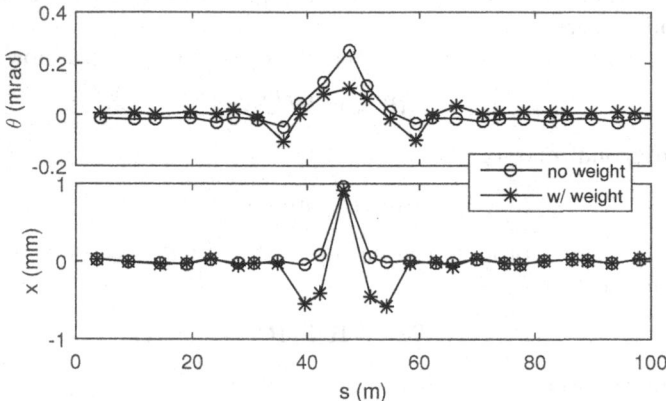

Figure 3.7 By reducing the weights on BPMs next to the target BPM, a local orbit bump could be created with lower corrector strengths. Top: corrector strengths with no weight on BPMs or with weights on the four BPMs next to the target BPM set to 0.2 (all others set to 1.0); Bottom: the resulting orbits by the two corrector solutions.

3.3.3 Other methods for global orbit correction

Before the orbit correction method based on SVD became popular, some other methods were used for global orbit correction. One method is called harmonics correction (or HARMON), which is aimed at the correction of the harmonic components of the closed orbit. This method is applicable to circular accelerators. Another method is called MICADO [5], which tries to solve the same least-square problem as defined in Eq. (3.8) with an iterative approach, using the most effective knob in each step. This method is applicable to both rings and one-pass systems.

The harmonic correction method is based on the observation that the closed orbit distortions are dominated by a few leading Fourier harmonics, as was shown in Eq. (2.15). The correctors can be changed in a certain pattern to target a specific orbit harmonic. According to Eq. (2.14), the corrector patterns on correctors $k = 1, 2, \cdots, N$ given by

$$\theta_{cn}(k) = \frac{\cos \frac{n\psi_k}{\nu}}{\sqrt{\beta_k}}, \quad \theta_{sn}(k) = \frac{\sin \frac{n\psi_k}{\nu}}{\sqrt{\beta_k}}, \tag{3.35}$$

will change the real and imaginary parts of the n'th orbit harmonic, f_n, respectively. The patterns affecting the real and imaginary parts can be considered independent knobs. With knobs that target a few harmonics around the betatron tune, the global orbit distortion can be substantially reduced.

The real and imaginary knobs for the harmonic $f_{[\nu]}$ are usually the most effective in changing the closed orbit, where $[\nu]$ is the closest integer to the

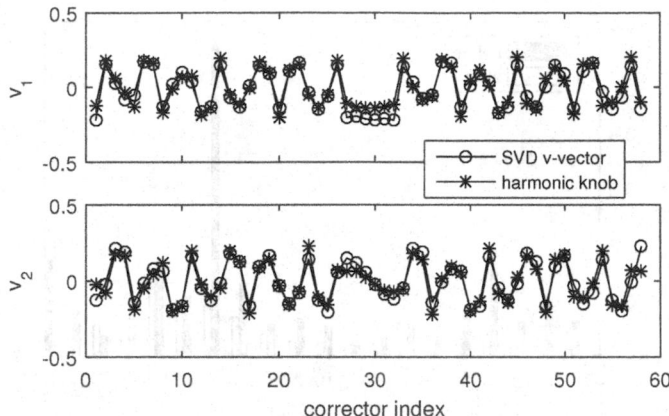

Figure 3.8 The **v**-vectors of the two leading SVD modes of the SPEAR3 horizontal orbit response matrix are compared to the real (top) and imaginary (bottom) knobs for the harmonic f_{14} (with $\nu_x = 14.106$).

betatron tune, ν. These knobs are closely related to the two leading SVD modes of the orbit response matrix. Figure 3.8 compares the v_1 and v_2 vectors of the horizontal orbit response matrix to the real and imaginary knobs for harmonic $f_{[\nu]}$, respectively, for the SPEAR3 storage ring. The correlation coefficients between the $v_{1,2}$ vectors and the corresponding harmonic knobs are 0.94 for both SV modes. An adjustment of the initial phase advance was made to align the v vectors and the harmonic knobs.

With a large number of distributed BPMs to measure the closed orbit, the orbit harmonics can be approximately computed with

$$F_n = \frac{v^2 f_n}{\nu^2 - n^2} = \frac{1}{M} \sum_{k=1}^{M} \frac{x_k e^{-in\psi_k/\nu}}{\sqrt{\beta_k}}, \tag{3.36}$$

where β_k and ψ_k are the beta function and the phase advance at BPM k, respectively. In a test example, the same closed orbit errors as in Figure 3.6 are corrected with the real and imaginary knobs of harmonic f_{14} to minimize the orbit distortion. The orbit harmonics before and after the orbit correction are shown in Figure 3.9. The rms orbit error becomes 0.90 mm after correction.

The MICADO method [5] for orbit correction has the same objective of minimizing the orbit errors in the least-square sense (i.e., Eq. (3.8)). It takes an iterative approach to find a solution for corrector changes. At the first step, it searches for the most effective orbit corrector knob by calculating the predicted residual vector for every knob after the orbit correction with that knob is done. At the second step, it looks for the most effective knob that, when combined with the first knob, reduces the predicted residual vector the most. Before the $(k+1)$'th step, it has identified k effective knobs. The order of the

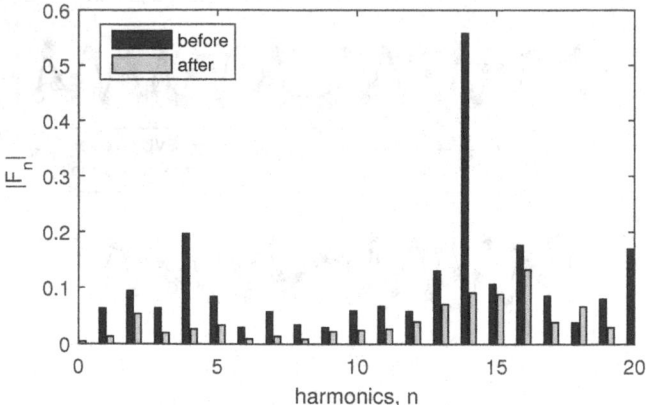

Figure 3.9 The horizontal orbit harmonics calculated with BPM readings before and after the same closed orbit errors as in Figure 3.6 are reduced with the two harmonic knobs for the integer stopband f_{14}.

knobs in vector $\boldsymbol{\theta}$ is shuffled to move all the selected knobs to the beginning. The columns in the orbit response matrix are rearranged correspondingly. At the $(k+1)$'th step, it looks for the most effective knob among the remaining $N-k$ knobs for orbit correction. Including the k selected knobs and one additional knob, corrector j, with $k+1 \leq j \leq N$, the solution to the $k+1$ included knobs is

$$\boldsymbol{\theta}_{k,j} = (\mathbf{R}_{1k,j}^T \mathbf{R}_{1k,j})^{-1} \mathbf{R}_{1k,j}^T \Delta \mathbf{x}, \qquad (3.37)$$

where $\boldsymbol{\theta}_{k,j} = (\theta_1, \theta_2, \cdots, \theta_k, \theta_j)^T$, and $\mathbf{R}_{1k,j}$ is the orbit response matrix for the $k+1$ included knobs. The vector $\boldsymbol{\theta}_{k,j}$ is used to compute the predicted residual vector. The knob j that results in the lowest objective function, χ^2, is then selected for the $(k+1)$'th step.

If the MICADO procedure is carried out for all corrector knobs, the solution is the same as using Eq. (3.16), and it would equally have difficulties with the near degeneracy in the orbit response matrix. The key point here is, however, to use only a few knobs for orbit correction. When only the most effective knobs are selected and the number of selected knobs is far fewer than N, the matrix $\mathbf{R}_{1k,j}^T \mathbf{R}_{1k,j}$ is usually invertible.

The MICADO method had been a useful global orbit correction method in the early days of synchrotrons, when orbit correctors were not as reliable and precisely controlled as today. Concentrating the correction on a few selected correctors with relatively large strengths helped achieve better reliability for the orbit correction system. With the modern technologies in corrector control, the SVD method is now the preferred method for global orbit correction in most cases.

3.3.4   Local orbit correction

As we have seen in the example in Section 3.3.2, local orbit bump can be created with the SVD method by specifying a target orbit that contains orbit changes only at the desired location(s). The desired orbit bump can involve multiple locations. For example, to make an angle bump at a location between two BPMs, the target beam positions on the two BPMs can be shifted in opposite directions. Even though all BPMs and corrector magnets are used in the calculation, typically only the correctors in the nearby region are changed. Changing the weights on the BPMs could help ease up the demand on the corrector strengths, at the cost of giving up the orbit control at some BPMs. This method should be able to meet the need of creating local orbit bumps whenever the orbit response matrix is available.

When an orbit bump is needed at a location without a BPM, the orbit response matrix can be extended to include an additional row for the target location. The elements for the row can be calculated with the lattice model. The SVD method can then be applied to calculate the required corrector pattern for the local bump.

The traditional method of creating local orbit bump does not rely on the orbit response matrix or the use of SVD. It is useful to understand the traditional approach as it would provide helpful insights. The requirement of a local bump is that the orbit change does not propagate outside of the last corrector. Suppose N correctors are involved in creating the local bump, this requirement amounts to

$$\sum_{i=1}^{N} \mathbf{M}(N|i) \begin{pmatrix} 0 \\ \theta_i \end{pmatrix} = \begin{pmatrix} 0 \\ 0 \end{pmatrix}, \tag{3.38}$$

where the summation is over all involved correctors, $i = 1$ for the first corrector, $i = N$ for the last corrector, and $\mathbf{M}(N|i)$ is the transfer matrix from corrector i to the exit face of corrector N.

The desired orbit bump can be a position change, an angle change, or both, at the target location, T. The target location needs not to be one of the correctors. In the general case, this can be specified with

$$\sum_{i=1}^{N_1} \mathbf{M}(T|i) \begin{pmatrix} 0 \\ \theta_i \end{pmatrix} = \begin{pmatrix} x \\ x' \end{pmatrix}, \tag{3.39}$$

where N_1 is the number of correctors before the target location, x and x' are the desired position and angle changes, respectively.

Conditions in Eqs. (3.38)-(3.39) contain four equations. In general, at least four correctors are required to satisfy these conditions, with at least two correctors before the target location. If the angle requirement in Eq. (3.39) can be ignored, three correctors would be enough. In this case, Eq. (3.38) leads to

$$\theta_1 \sqrt{\beta_1} \sin \psi_{31} + \theta_2 \sqrt{\beta_2} \sin \psi_{32} = 0, \tag{3.40a}$$

$$\theta_2 \sqrt{\beta_2} \sin \psi_{21} + \theta_3 \sqrt{\beta_3} \sin \psi_{31} = 0. \tag{3.40b}$$

For the simplicity of discussion, we assume the target location is at corrector 2. The position condition in Eq. (3.39) becomes

$$x = \theta_1 \sqrt{\beta_1 \beta_2} \sin \psi_{21}. \tag{3.41}$$

Therefore, to make a large bump, it is desirable to have ψ_{21} close to $\frac{\pi}{2} + k\pi$, with integer k. The kicks provided by correctors 2 and 3 are needed to eliminate the angle and position coordinate changes after corrector 3. If both ψ_{21} and ψ_{32} are equal to $\frac{\pi}{2}$ modulo π, then ψ_{31} is a multiple of π, and the strength for corrector 2 is zero.

Linear optics measurement and correction - I

The linear optics of an accelerator beam line is of fundamental importance - after all, the main purpose of the beam line lattice is to achieve the desired linear optics. In a one-pass lattice, the linear optics is designed to preserve the beam quality, avoid beam loss, and deliver certain beam characteristics at selected locations. In a circular accelerator, the linear optics not only determines the transverse distributions of the beam, but also has significant impact to its control and stability. A storage ring with large linear optics errors, as it often happens during the commissioning stage of a new ring, can fail to store the beam. In an electron storage ring, the linear optics also determines the equilibrium distribution of the beam.

There are many error sources that can cause deviations of the linear optics from the design. Quadrupole magnets could have strength errors due to manufacturing errors, magnetic field calibration errors, or power supply regulation errors. Horizontal orbit offsets in sextupole magnets, which can be caused by magnet misalignment, introduce quadrupole components from the feed-down effect. Other types of magnets, such as dipole or sextupole magnets, could have random quadrupole errors due to manufacturing errors. Inaccurate lattice modeling could introduce discrepancies between the model and the real machine. Insertion devices could contribute to linear optics errors through the quadrupole components in their residual field integrals or the dynamical effects that arise from the transverse field roll-off and the sinusoidal trajectories [103]. Impedances could also cause optics errors for an intense beam.

With linear optics errors, the betatron tunes are usually shifted from the design values. The lattice functions lose the periodicity. Because of the beta beating and betatron phase beating, the sextupole cancellation scheme could lose effectiveness, causing degradation of the nonlinear beam dynamics performance. Correction of the linear optics toward the design has many benefits.

Similar to orbit correction, correction of linear optics requires a correction target, measurements that characterize the optics errors, and knobs to compensate the errors. The correction target is typically the design optics, which is often represented by a lattice model with which the linear optics functions, such as the Courant-Snyder parameters, the betatron phase advances, and the dispersion functions, can be computed.

To correct the linear optics errors, measurements must be conducted to sample the linear optics and determine the errors. Magnetic field measurement on an operating accelerator is not realistic; even if it can be done, it cannot resolve all differences between the machine and the design as some differences are visible only to the beam (e.g., the dynamic ID effect and the impedance effect) and some differences come from the model (e.g., hard-edge approximation of the magnetic field profiles). Beam-based measurements can reveal the linear optics of the machine as the beam experiences. Using information derived from beam-based measurements to correct the linear optics can restore the optics condition for the beam.

In the case of orbit correction, the errors that need correction can be readily determined since the orbit is measured directly with BPMs. The linear optics, however, cannot be directly obtained. Indirect measurements are generally necessary to determine the linear optics. Processing data taken in these measurements to extract the optics errors and to derive the required knob changes for correction is a fundamental challenge in linear optics correction. Closed-orbit response matrix [77, 76, 44, 25, 21, 102] and turn-by-turn BPM data [15, 19, 119, 118, 58, 61, 116, 125, 117] are two basic types of measurements that are used for linear optics measurement and correction. The measured dispersion function is often used as supplemental data.

Beta functions in circular accelerators can also be determined by measuring tune shifts due to quadrupole modulation [48]. In this method, the strength of a quadrupole is varied and the corresponding betatron tune shifts are measured. Knowing the length of the quadrupole magnet and the current to gradient conversion rate, the beta function at the location of the quadrupole can be calculated with Eq. (2.22). The accuracy of the method is affected by the hysteresis of the quadrupole magnet and calibration errors.

Because of the large number of potential error sources, it is not feasible to correct all linear optics errors at the error sources. Instead, quadrupole magnets are used as knobs to compensate the linear optics errors globally. In early accelerators, quadrupole magnets are often powered in series. The strengths of individual quadrupoles can be changed with shunt resistors that divert a fraction of the current. Quadrupole magnets in newer machines are often individually powered. Their strengths can be changed by adjusting the setpoints of the corresponding power supplies. Some machines have quadrupole correctors beside the main quadrupole magnets.

In this chapter we will discuss the optics correction method that uses orbit response matrix [102] as the input data. The methods that are based on turn-by-turn BPM data [58, 61, 116, 125] will be discussed in the next chapter.

4.1 BEAM MEASUREMENTS FOR LINEAR OPTICS

4.1.1 Sampling linear optics with transverse beam profile

The primary goal of the linear optics of a magnet lattice is to keep the beam focused in both transverse planes, i.e., to preserve small beam sizes through the beam line. Linear optics concerns the transformation of the transverse beam profile from one location to another. The transverse distribution of a particle beam is characterized by its second order moment matrix, Σ, (see Eq. (1.80)). The propagation of the transverse distribution through the lattice is specified by the transfer matrix between locations. The Σ-matrices at two locations, points 1 and 2, are related via

$$\Sigma_2 = \mathbf{M}_{21}\Sigma_1\mathbf{M}_{21}^T, \tag{4.1}$$

where $\mathbf{M}_{21}$ is the transfer matrix from point 1 to point 2. Transfer matrices are fundamental representation of the linear optics of a lattice. The Courant-Snyder optics functions (α, β, and γ) and the betatron phase advances are an equivalent form of the linear optics description as they can be used to construct the transfer matrix.

In principle, the linear optics properties can be determined by beam profile measurements, using Eq. (4.1). This is the basis of the quadrupole scan method for emittance measurement, which is widely used in linacs or transport lines. In this method, the transverse beam profile is measured with a screen or a wire scanner while the strength of an upstream quadrupole magnet is varied. The quadrupole is separated from the screen by a drift space of length L. Letting the entrance of the quadrupole be point 1 and the screen be point 2, the transfer matrix from point 1 to 2 can be readily calculated as a function of the quadrupole strength. The beam distribution at point 1 can be given in terms of the beam emittance and the Courant-Snyder parameters by

$$\Sigma_1 = \begin{pmatrix} \sigma_1^2 & \sigma_{12} \\ \sigma_{12} & \sigma_{1'}^2 \end{pmatrix} = \epsilon \begin{pmatrix} \beta_1 & -\alpha_1 \\ -\alpha_1 & \gamma_1 \end{pmatrix}. \tag{4.2}$$

Σ_1 is not affected by the quadrupole strength. From Eq. (4.2), the measured rms beam size at point 2, σ_2^2, is related to the integrated quadrupole gradient, $k = K_1 L_q$ (with quadrupole gradient K_1 and length L_q), and the beam parameters at point 1 by

$$\sigma_2^2 = (L^2\sigma_1^2)k^2 + (2L\sigma_1^2 + 2L^2\sigma_{12})k + (\sigma_1^2 + 2L\sigma_{12} + L^2\sigma_{1'}^2). \tag{4.3}$$

By fitting σ_2^2 to a quadratic function of k and identifying the coefficients, the second order moments at point 1 can be determined, which in turn can be used to calculate the emittance and Courant-Snyder parameters.

The quadrupole scan method can determine the beam profile and optics functions at one location of a one-pass system. The measured optics functions can be used to calculate the required quadrupole strength adjustments to meet the desired optics matching conditions. It is not suitable for rings because in a ring the beam profile at the entrance of the quadrupole is changed by its own strength and is thus not fixed. Equilibrium beam sizes in electron storage rings can be measured with pinhole cameras that image the beam profile in a dipole magnet through synchrotron radiation, which can be used to determine the beta functions at the radiation source points (up to a scaling constant). However, because beam profile measurements are available only at few locations, this method cannot be used to measure the global linear optics.

4.1.2 Sampling linear optics with beam orbit

Another way to sample the linear optics is through the beam trajectory or orbit. The transfer matrix directly relates the phase space coordinates of a single particle at two locations through

$$\mathbf{X}_2 = \mathbf{M}_{21}\mathbf{X}_1, \tag{4.4}$$

where $\mathbf{X} = (x, x')^T$ or $(y, y')^T$. The BPMs measure the average position of all particles in the beam, i.e., the beam center.

When the beam is kicked, in some cases all particles in the beam move with the beam center, without changing the phase space distribution around the beam center. In such cases, the beam center behaves like a single particle. For example, one such case is when the strength of a corrector magnet is changed in a transport line or a storage ring.

The beam may also be kicked by a kicker or a pinger in a storage ring during only one pass. After the kick all particles in the beam will start to oscillate. Because the betatron tunes of the particles may differ, due to amplitude or chromatic detuning, the particles will gradually move out of phase. The beam center will move toward the closed orbit as particles populate along ellipses centered on the closed orbit, even though the individual particles are still oscillating with large amplitude. This phenomenon is called decoherence. The beam center for a beam undergoing decoherence does not behave like a single particle. However, the decay of the apparent motion of the beam center takes tens to hundreds of turns. Orbit data taken during this period can still be seen as representing the motion of a single particle. Therefore, orbit data taken with both orbit corrector changes or one-pass kicks can be used to sample the linear optics.

Since BPMs can only directly measure the position coordinates, x or y, not the angle coordinates, Eq. (4.4) usually cannot be used to measure the transfer matrix. However, if two BPMs are separated by a beam line of which the transfer matrix is known, then the angle coordinates at the two BPMs can be calculated using the position coordinates and the transfer matrix between the two BPMs,

$$x_1' = \frac{x_2 - M_{11}x_1}{M_{12}}, \quad x_2' = M_{21}x_1 + M_{22}x_1', \tag{4.5}$$

where M_{11}, M_{12}, M_{21}, and M_{22} are the four elements of the transfer matrix from BPM 1 to BPM 2. For the special case when the beam line between the two BPMs is a drift space, the transfer matrix is known with a high accuracy. In this case, we have

$$x_1' = x_2' = \frac{x_2 - x_1}{L}, \tag{4.6}$$

where L is the distance between the BPMs.

If the phase space coordinates can be determined at two locations (with a pair of BPMs for each location) using the above method, the transfer matrix between the two locations can be determined from Eq. (4.4). In principle, only two different orbits are needed to determine the four matrix elements. In reality, more orbit data can be used to achieve a higher accuracy under random noise in the data. The orbit data with N orbit measurements can be arranged in the form of

$$\mathbf{X}_1 = \begin{pmatrix} x_1(1) & x_1(2) & \cdots x_1(N) \\ x_1'(1) & x_1'(2) & \cdots x_1'(N) \end{pmatrix}, \quad \mathbf{X}_2 = \begin{pmatrix} x_2(1) & x_2(2) & \cdots x_2(N) \\ x_2'(1) & x_2'(2) & \cdots x_2'(N) \end{pmatrix}, \tag{4.7}$$

for BPMs 1 and 2, respectively. Using the least-square method to fit the matrix elements, it is straightforward to show that the transfer matrix can be found with

$$\mathbf{M}_{21} = \mathbf{X}_2 \mathbf{X}_1^T (\mathbf{X}_1 \mathbf{X}_1^T)^{-1}. \tag{4.8}$$

Because of random errors, the transfer matrix obtained with Eq. (4.8) is not strictly symplectic. An equivalent symplectic transfer matrix can be obtained from $\mathbf{M}$ using a procedure given in Ref. [45]. The symplectic matrix can also be obtained by fitting a set of parameters from which the symplectic matrix is constructed. The Courant-Snyder parameters can be used for the 2×2 transfer matrix. A 4×4 transfer matrix, can be constructed with 10 parameters using the procedure described in Eqs. (2.59-2.63) [105, 61].

In the case of storage rings, phase space orbit data at one location over successive turns can be used to fit the one-turn transfer matrix with the above method. The orbit for turn n can be seen as the data from BPM 1 and the orbit for turn $n + 1$, the data from BPM 2, with $n = 1, 2, \cdots, N - 1$. In a storage ring light source, there is usually one pair of BPMs separated by a long drift space for each period (where the drift space is used to house insertion devices). Therefore, the one-turn transfer matrix at the straight section of each period and the transfer matrix across the period can be determined. The transfer matrices can be used to calculate the Courant-Snyder parameters and phase advances.

The method of measuring the transfer matrix with pairs of BPMs separated by drift spaces is applicable only to a limited number of locations and may not provide enough sampling for global optics correction. Even for storage ring light sources, the BPMs in the straight sections account for a small fraction of BPMs. The optics information for the BPMs in the arcs is missing. Other methods that make use of the orbit data at all BPMs are needed in order to measure and correct the linear optics throughout the lattice.

Depending on the type of orbit data in use, these methods can be put in two categories: methods of turn-by-turn (or pass-by-pass for one-pass systems) BPM data and the orbit response methods.

For the turn-by-turn (TbT) methods, BPM data are taken with the beam shifted away from the normal orbit. The beam orbit deviations propagate through the lattice to downstream locations and are recorded by the BPMs. Because the deviations at different locations are related through the linear optics (Eq. (4.4)), it is possible to extract optics information from the BPM data. In a circular accelerator, after the beam motion is excited, the subsequent free betatron oscillation moves the orbit along the phase space ellipse according to

$$x(n) = \sqrt{2\beta J} \cos(2\pi\nu n + \xi), \tag{4.9a}$$

$$x'(n) = -\sqrt{\frac{2J}{\beta}} \sin(2\pi\nu + \xi) - \frac{\alpha}{\beta} x(n), \tag{4.9b}$$

where ν is the betatron tune, n is the turn number, and J and ξ are the action and phase variables given by the initial condition, respectively. Because the fractional betatron tune is typically not equal to a low order rational number, within a few hundreds of turns, the beam will spread out over the entire ellipse and hence sample the phase space from all angles, as illustrated in Figure 4.1 (a).

In a one-pass system, the sampling of the betatron phase space can be achieved by scanning two upstream corrector magnets [33, 128]. The two

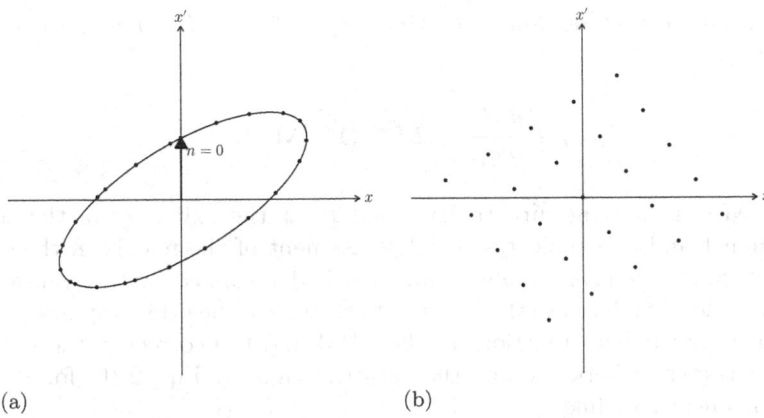

Figure 4.1 Sampling the betatron phase space by kicking the beam away from the origin of the phase space. (a): beam kicked in a ring; (b): scanning two upstream correctors in a one-pass system.

correctors will shift the position and angle coordinates of the beam on downstream BPMs. Varying the kicks by the two correctors, the beam orbit can scan the entire phase space. This is as illustrated in Figure 4.1 (b).

Turn-by-turn or pass-by-pass BPM data can be used to determine the optics errors in the lattice. This can be done by comparing the BPM data to tracking data produced by a lattice model and using a fitting method to adjust the quadrupole parameters in the model to minimize the differences between the measured and tracking data [61]. This method is applicable to both rings and one-pass systems. Turn-by-turn BPM data taken from a storage ring contain temporal oscillations of the beam position. The oscillation signals on different BPMs are correlated but with different amplitudes and phases. The differences reflect the beta functions and betatron phase advances. There are multiple methods to extract the optics functions by analyzing turn-by-turn BPM data. These methods will be discussed in the next chapter.

The second category of linear optics sampling with beam orbits uses the orbit responses. In a one-pass system, the orbit response for a thin-element corrector is simply the $(1, 2)$ element of the transfer matrix, i.e.,

$$R_{ij} \equiv \frac{dx_i}{d\theta_j} = M_{12}^{(ij)}, \tag{4.10}$$

where R_{ij} is defined as the orbit response at BPM i for a kick at corrector j, and $\mathbf{M}^{(ij)}$ is the transfer matrix from the corrector to the monitor. The orbit response in a one-pass system is non-zero only if the monitor is downstream of the corrector magnet.

In a circular accelerator, the orbit response at a BPM is given by (see Eq. (2.8))

$$R_{ij} \equiv \frac{dx_i^{(c)}}{d\theta_j} = [\mathbf{M}^{(ij)}(\mathbf{I} - \mathbf{M}^{(j)})^{-1}]_{12}, \tag{4.11}$$

where $\mathbf{M}^{(j)}$ is the one-turn transfer matrix at the exit edge of the corrector magnet and $[\cdot]_{12}$ indicates the $(1, 2)$ element of the matrix in the square bracket. Since the orbit response in a ring is determined by the transfer matrix, it is closely related to the linear optics. In fact, the orbit response is given explicitly by the beta functions at the BPM and the corrector, the betatron phase advance in between, and the betatron tune by Eq. (2.10) for the case without linear coupling.

Each column of the orbit response matrix is a differential orbit from the original orbit. It corresponds to one sampling point of the betatron phase space in the same sense as the orbit shift from the origin for the turn-by-turn or pass-by-pass BPM data (see Figure 4.1). There is no difference between the pass-by-pass BPM data with corrector scans and the orbit response data (or more accurately, trajectory response) for one-pass systems. For the case of rings, the only difference between turn-by-turn BPM data and closed orbit responses is that the closed orbit is abruptly changed at the location of the corrector by the kick it applies. At all other locations, the closed orbit response represents the free betatron motion. The closed orbit measurement is typically much more accurate than the turn-by-turn orbit measurement as the former employs averaging of beam signals over many turns. However, turn-by-turn BPM data of hundreds or more turns could be collected. By properly processing the multi-turn BPM data, the same precision could be achieved with turn-by-turn BPM data as the closed orbit data in linear optics measurement.

With the measured orbit data that sample the betatron phase space, the next step is extract the linear optics from the data. Typically this is done by fitting a lattice model with the data. In the next section we will describe the fitting method for the orbit response matrix.

4.2 FITTING ORBIT RESPONSE MATRIX TO LATTICE MODEL

There are two goals in analyzing the orbit data for linear optics: optics measurement and optics correction. The linear optics in a storage ring is often described by the beta functions and the betatron phase advances. The purpose of optics measurement is to derive these functions from the data. For optics correction, the goal is to compensate the errors in the linear optics such that the linear optics is as close to the design optics as possible.

It would seem natural to accomplish these two goals in two steps: first derive the optics functions with optics measurement, then use the optics functions to determine the required changes to the quadrupoles for optics correction. This is actually the case for several methods that use the turn-by-turn

BPM data (to be discussed in the next chapter). Given the connection between the orbit responses and the optics functions (see Eq. (2.10)), it is possible to derive the optics functions directly from the orbit response data, for example, by fitting the data for the beta functions and the phase advances at all BPMs and correctors. However, this is an unnecessary step if the final goal is optics correction and it could introduce errors. The typical approach of extracting linear optics information from the orbit response matrix data is to fit for the quadrupole errors with a lattice model. Through fitting, the lattice model is calibrated, which means that it is made consistent with the measurements. It can then be used to calculate the optics functions or other lattice parameters.

When there are linear optics errors, the dispersion function will also be distorted. Measuring the dispersion function and including it in the fitting will help determine the optics errors and make the fitting results more effective for restoring the design dispersion.

The method of measuring and correcting storage ring linear optics by fitting orbit response matrix data and dispersion data is commonly referred to as Linear Optics from Closed Orbit (LOCO). It was first successfully demonstrated on the NSLS rings [102]. The method was later implemented in an easy-to-use program [104, 95] and its fitting algorithm was updated to handle the degeneracy issue [49, 52, 50]. It has become a widely used tool, especially in the synchrotron light source community.

4.2.1 Measured orbit response matrix

Because quadrupole magnets affect the optics functions in both transverse planes, the orbit response matrices in both planes are used to fit the quadrupole errors. The orbit responses in both planes can be arranged in one matrix

$$\mathbf{R} = \begin{pmatrix} \mathbf{R}_{xx} & \mathbf{R}_{xy} \\ \mathbf{R}_{yx} & \mathbf{R}_{yy} \end{pmatrix}, \tag{4.12}$$

where the first x or y in the subscript indicates the plane of the BPMs and the second indicates the plane of the orbit correctors. For example, $\mathbf{R}_{xx}$ contains the horizontal orbit responses of the horizontal correctors, and $\mathbf{R}_{xy}$ contains the horizontal orbit responses of the vertical correctors,

$$\mathbf{R}_{xx} = \frac{d\mathbf{x}}{d\boldsymbol{\theta}_x}, \quad \mathbf{R}_{xy} = \frac{d\mathbf{x}}{d\boldsymbol{\theta}_y}, \tag{4.13}$$

where $\mathbf{x}$ is a column vector of all horizontal BPM readings, $\boldsymbol{\theta}_{x,y}$ are the column vectors for the horizontal and vertical kick angles, respectively, and similarly,

$$\mathbf{R}_{yx} = \frac{d\mathbf{y}}{d\boldsymbol{\theta}_x}, \quad \mathbf{R}_{yy} = \frac{d\mathbf{y}}{d\boldsymbol{\theta}_y}. \tag{4.14}$$

The off-diagonal blocks $\mathbf{R}_{xy}$ and $\mathbf{R}_{yx}$ are non-zero if there is linear coupling between the x and y planes or rolls of BPMs or correctors about the s-axis.

Typically, a BPM measures the horizontal and vertical beam positions simultaneously. Suppose there are M BPMs, N_x horizontal correctors, and N_y vertical correctors, the dimension of $\mathbf{R}_{xx}$ and $\mathbf{R}_{yx}$ is $M \times N_x$ and the dimension of $\mathbf{R}_{xy}$ and $\mathbf{R}_{yy}$ is $M \times N_y$. The dimension of $\mathbf{R}$ is thus $2M \times (N_x + N_y)$.

The orbit response of an orbit corrector is measured on the machine by stepping the strength of the corrector and measuring the orbit shift. Because of the presence of nonlinear magnets in the lattice, the orbit response is nonlinear to some extent. The impact of the nonlinearity to the accuracy of the orbit response matrix can be minimized by using the bipolar scheme in the measurement, in which the corrector strength is changed in both the negative and positive directions by the same step size, $\Delta\theta$. The orbit responses are then calculated with

$$\mathbf{R}_x = \frac{\mathbf{x}_+ - \mathbf{x}_-}{2\Delta\theta}, \quad \mathbf{R}_y = \frac{\mathbf{y}_+ - \mathbf{y}_-}{2\Delta\theta}, \tag{4.15}$$

where subscripts $+$ and $-$ indicate the orbits for the positive and negative steps, respectively. The step size of the corrector change is preferred to be small in order to reduce the impact of the nonlinearity. On the other hand, a large orbit shift relative to the BPM noise is desired for high data precision. A good trade-off between the two conflicting requirements may depend on the specific machine. Corrector step changes that causes orbit shifts of 1-2 mm are usually a reasonable choice for a typical third generation light source.

In the measurement of the orbit response matrix, orbit correctors are used to change the beam orbit and the BPMs are used to detect the orbit changes. In reality, the correctors and BPMs can both have calibration errors. They could also have rotations about the s-axis. These errors will cause the measured orbit response matrix to differ from the actual response matrix and hence need to be accounted for. The actual kicks by a corrector are related to the apparent kick values by

$$\begin{pmatrix} \theta_x \\ \theta_y \end{pmatrix} = \begin{pmatrix} \cos\phi & \sin\phi \\ -\sin\phi & \cos\phi \end{pmatrix} \begin{pmatrix} k_x \tilde{\theta}_x \\ k_y \tilde{\theta}_y \end{pmatrix}, \tag{4.16}$$

where $k_{x,y}$ are the horizontal and vertical gains of the corrector kicks, ϕ is the rotation about the s-axis, and $\tilde{\theta}$ and θ represent the actual and apparent kicks, respectively.

In addition to the calibration and rotation errors, BPMs could have another type of errors that arises from the deviation of the button positions from the ideal configuration. With these errors, the actual beam positions ($\tilde{x}$ and $\tilde{y}$) and the apparent positions (x and y) reported by the BPMs are related via [102]

$$\begin{pmatrix} x \\ y \end{pmatrix} = \frac{1}{\sqrt{1 - C^2}} \begin{pmatrix} \cos\phi & \sin\phi \\ -\sin\phi & \cos\phi \end{pmatrix} \begin{pmatrix} 1 & C \\ C & 1 \end{pmatrix} \begin{pmatrix} g_x \tilde{x} \\ g_y \tilde{y} \end{pmatrix}, \tag{4.17}$$

where $g_{x,y}$ are the horizontal and vertical BPM gains, respectively, ϕ is the rotation angle about the s-axis, and C is called the crunch coefficient, which represents the effect of button configuration distortion. Eq. (4.17) can be written in an equivalent form

$$\begin{pmatrix} x \\ y \end{pmatrix} = \begin{pmatrix} g_x & c_x \\ c_y & g_y \end{pmatrix} \begin{pmatrix} \tilde{x} \\ \tilde{y} \end{pmatrix}, \tag{4.18}$$

where the g_x and g_y are redefined BPM gains and c_x and c_y are the coupling coefficients.

For the purpose of extracting optics errors in the machine, we should compare the actual orbit response matrix to the model orbit response matrix. To calculate the actual orbit response matrix from the raw measured data, error parameters (gains, roll, and crunch) of the correctors and BPMs are used in Eqs. (4.16) and (4.17). In general, these parameters are not known in advance and need to be included as fitting parameters.

The dispersion function is often included in the orbit response matrix fitting. The dispersion is measured by shifting the RF frequency by a small amount, Δf, and observing the closed orbit changes. In a storage ring, when the RF frequency is shifted, the beam momentum has to change in order for the beam to stay synchronous with the RF cavity, with the momentum deviation given by

$$\delta = -\frac{\Delta f}{\alpha_c f_{\mathrm{rf}}}, \tag{4.19}$$

where f_{rf} is the RF frequency. The closed orbit dependence on beam energy also has nonlinear terms. The bipolar scheme with an appropriate step size can help minimize the impact of the nonlinear dispersion to the measurement accuracy. The measured horizontal and vertical dispersion functions are

$$D_x = -\alpha_c f_{\mathrm{rf}} \frac{\mathbf{x}_+ - \mathbf{x}_-}{2\Delta f}, \quad D_y = -\alpha_c f_{\mathrm{rf}} \frac{\mathbf{y}_+ - \mathbf{y}_-}{2\Delta f}, \tag{4.20}$$

where subscripts $+$ and $-$ indicate orbits measured with the positive or negative frequency shifts, respectively. BPM gains, rolls, and geometric distortion errors also affect the dispersion measurements.

4.2.2 Model orbit response matrix

With a lattice model, the closed orbit can be numerically computed by looking for the orbit that satisfies the fixed-point condition, Eq. (2.6). The orbit response of an orbit corrector can be calculated using the bipolar scheme in the same manner as in the measurement. It is desirable to also choose the same corrector step size as in the measurement.

If the corrector magnet is located in a dispersive region, a change of the corrector kick will change the path length of the closed-orbit, as described in

Eq. (2.50). In a storage ring the path length of the beam orbit is at a fixed ratio with the RF frequency. Therefore, after the corrector kick the beam energy will change by

$$\frac{\Delta E}{E} = \frac{D\theta}{\alpha_c C},$$ (4.21)

in order for the beam to stay synchronous with the RF cavity. The energy shift will cause orbit shifts at dispersive locations and will be reflected in the measured orbit response matrix.

The calculation of the orbit response matrix from the lattice should include the effect of the energy shifts due to horizontal corrector kicks. If the closed orbit is calculated considering the transverse planes only (i.e., satisfying the closed orbit condition only in (x, x', y, y') coordinates), the beam energy is equal to the design energy and is constant. Thus the energy shift is not included in the orbit response matrix. In this case, an additional term should be added to each element of the horizontal block $\mathbf{R}_{xx}$,

$$\mathbf{R}_{ij} \rightarrow \mathbf{R}_{ij} + \frac{D_i D_j}{\alpha_c C},$$ (4.22)

for the element corresponding to BPM i and corrector j.

If the closed orbit is calculated for the 6-dimensional coordinates with the requirement $\Delta z = 0$, the path length of such a closed orbit is not changed by the corrector kick as it will automatically include the proper energy shift.

4.2.3   Least-square fitting setup

The differences between the measured and model orbit response matrices are due to the linear optics errors in the actual machine, systematic errors in the measurements (such as BPM and corrector gain errors), and random noise in the BPM readings. The linear optics and systematic measurement errors could be determined by adjusting the corresponding parameters in the lattice model or the measurements to minimize the differences. The lattice parameters are the quadrupole gradients. The differences between the measured and model orbit response matrices can be characterized by

$$\chi^2 = \sum_{ij} \frac{1}{\sigma_{ij}^2} (R_{ij}^{\text{meas}} - R_{ij}^{\text{model}})^2,$$ (4.23)

where σ_{ij} is the rms noise level of the matrix element R_{ij}^{meas}. The corrector current is often very precisely regulated and its noise can be neglected. Therefore, the element noise sigma is

$$\sigma_{ij} = \frac{\sigma_i}{\theta_j},$$ (4.24)

where σ_i is the noise sigma for the monitor i and θ_j is the kick angle step change for the orbit response measurement of corrector j. It may be convenient to define the step kick angle change as the unit angle. Then in Eq. (4.23) σ_{ij} is replaced with σ_i and the matrix elements are interpreted as orbit shifts in length unit (mm or m). The function χ^2 depends on the lattice parameters through $\mathbf{R}^{\text{model}}$ and on the measurement parameters (gains and rolls) through $\mathbf{R}^{\text{meas}}$. The objective of the least-square fitting is to minimize χ^2.

When the dispersion functions are included in the fitting problem, the objective function includes additional terms

$$\chi^2 = \sum_{ij} \frac{1}{\sigma_i^2}(R_{ij}^{\text{meas}} - R_{ij}^{\text{model}})^2 + \left(\frac{\alpha_c f_{\text{rf}}}{\Delta f}\right)^2 \sum_i \frac{1}{\sigma_i^2}(D_i^{\text{meas}} - D_i^{\text{model}})^2,$$

(4.25)

where i goes from 1 to $2M$ to include both horizontal and vertical dispersion functions. The dispersion function terms can be seen as one column of the orbit response matrix.

The terms in Eq. (4.23) can be arranged by defining the residual vector, $\mathbf{r}$, with its k-th element being

$$r_k = \frac{1}{\sigma_i}(R_{ij}^{\text{meas}} - R_{ij}^{\text{model}}), \quad k = 2(j-1)M + i,$$

(4.26)

where $2M$ is the number of rows in the orbit response matrix. Including the dispersion functions, the length of the residual vector is $2M(N_x + N_y + 1)$. The objective function can thus be written in the standard form

$$f(\mathbf{p}) \equiv \chi^2 = \mathbf{r}^T \mathbf{r},$$

(4.27)

where the parameter vector $\mathbf{p}$ contains all of the fitting parameters.

The fitting parameters may include N_q quadrupole parameters, N_{sq} skew quadrupole parameters, horizontal and vertical BPM gains and coupling coefficients ($4M$ BPM parameters in total), horizontal corrector gains and rolls ($2N_x$), vertical corrector gains and rolls ($2N_y$). There are a total of $N_q + N_{sq} + 4M + 2N_x + 2N_y$ fitting parameters. The skew quadrupole parameters, BPM roll and shape distortion, and corrector rolls are used to account for the cross-plane coupling (the off-diagonal blocks of the response matrix). To first order, the linear optics affects only the diagonal blocks. If we are concerned only of the linear optics, then the skew quadrupoles and BPM and corrector rolls can be left out. In this case, there are $N_q + 2M + N_x + N_y$ fitting parameters.

4.2.4   Gauss-Newton method

The function $f(\mathbf{p})$ is generally nonlinear with respect to $\mathbf{p}$. An iterative approach is usually applied to find the solution that minimizes the objective function. The initial solution, $\mathbf{p}_0$, may be given by the ideal values of the fitting parameters. The initial quadrupole strengths can be the design values.

The initial BPM and corrector gains are all set to unity. The initial BPM and corrector coupling coefficients are set to zero.

Suppose the residual vector corresponding to $\mathbf{p}_0$ is $\mathbf{r}_0$, the residual vector after the fitting parameters are changed to $\mathbf{p} = \mathbf{p}_0 + \Delta\mathbf{p}$ can be linearly expanded to give

$$\mathbf{r} = \mathbf{r}_0 + \mathbf{J}\Delta\mathbf{p}, \quad \text{with} \quad J_{ij} = \frac{\partial r_i}{\partial p_j}, \tag{4.28}$$

where $\mathbf{J}$ is the Jacobian matrix of the residual vector with respect to the parameter vector $\mathbf{p}$, whose matrix elements are as defined in the above. Each column of $\mathbf{J}$ represents the differential impact of the corresponding fitting parameter to the residual vector. The objective function can also be expanded around $\mathbf{p}_0$, which is approximately given by

$$f(\mathbf{p}) \approx f(\mathbf{p}_0) + 2\mathbf{r}_0^T \mathbf{J}\Delta\mathbf{p} + \Delta\mathbf{p}^T \mathbf{J}^T \mathbf{J}\Delta\mathbf{p}, \tag{4.29}$$

where we neglected the quadratic and higher order dependence of $\mathbf{r}$ on the fitting parameters. At the present solution, $\mathbf{p}_0$, the gradient is $2\mathbf{J}^T\mathbf{r}_0$ and the approximate Hessian matrix is given by $2\mathbf{J}^T\mathbf{J}$. Because the gradient of the objective function, $\nabla_\mathbf{p} f(\mathbf{p}) = 2\mathbf{J}^T\mathbf{r}_0 + 2\mathbf{J}^T\mathbf{J}\Delta\mathbf{p}$, is zero at a minimum of $f(\mathbf{p})$, the solution for the step change toward the minimum from the present solution is found to be

$$\Delta\mathbf{p} = -(\mathbf{J}^T\mathbf{J})^{-1}\mathbf{J}^T\mathbf{r}_0. \tag{4.30}$$

This solution is the same as Eq. (3.16), here with the Jacobian matrix in place of the orbit response matrix. In fact, the orbit response matrix is the Jacobian matrix of the beam orbit with respect to the corrector strengths. Solving least-square problems by iteratively applying Eq. (4.30) is called the Gauss-Newton method.

Similar to the orbit correction case, Eq. (4.30) is applicable only if the square matrix $\mathbf{J}^T\mathbf{J}$ is invertible. This condition is equivalent to require that all singular values of $\mathbf{J}$ be nonzero, with the SVD of the Jacobian matrix given in the form, $\mathbf{J} = \mathbf{U}\mathbf{S}\mathbf{V}^T$. However, in general, this condition is not met. Consider a simple case without roll and crunch errors to BPMs or correctors. If the residual vector contains only the orbit response matrix terms (i.e., not including the dispersion terms), the Jacobian matrix is degenerate with two zero singular values, one for each transverse plane. The degeneracy comes from the fact that if all corrector gains and all BPM gains of the same plane are raised by a common factor, the measured orbit response matrix does not change (since in $R_{ij} = \Delta x_i / \Delta\theta_j$ both the nominator and denominator change by the same ratio). Including the dispersion terms in the residual vector alleviates the degeneracy because these terms are not dependent on the corrector gains and RF frequency measurement is very accurate. It is more helpful in the horizontal plane as the orbit shift due to horizontal dispersion is much larger than the spurious vertical dispersion.

Because of the degeneracy in the Jacobian matrix, it is necessary to compute the pseudo-inverse of the square matrix $\mathbf{J}^T\mathbf{J}$. This can be done in the same fashion as was done in Chapter 3 for the orbit correction case, which results in the solution of $\Delta\mathbf{p}$ in the form of

$$\Delta\mathbf{p} = -\sum_{i=1}^{N_{\text{th}}} \frac{1}{s_i}\mathbf{v}_i(\mathbf{u}_i^T\mathbf{r}_0), \qquad (4.31)$$

where $\mathbf{u}_i$ and $\mathbf{v}_i$ are column vectors in matrices $\mathbf{U}$ and $\mathbf{V}$ for the i'th singular value, respectively, and N_{th} stands for the number of singular values to be kept in the calculation of the pseudo-inverse matrix. In orbit response matrix fitting, the number of rows of the Jacobian matrix is often significantly larger than the number of columns (i.e., the number of fitting parameters). For example, for a ring with 180 BPMs, 180 horizontal correctors, and 180 vertical correctors, the dimension of the full orbit response matrix, and hence the number of rows, is 360×360. The $\mathbf{U}$-matrix in the SVD of the Jacobian matrix would have $360^4 = 1.68 \times 10^{10}$ elements. The space needed to store such a matrix exceeds the memory size of an ordinary computer. Fortunately, in Eq. (4.31) we only need the columns of the $\mathbf{U}$-matrix that correspond to the non-zero singular values. This can be obtained without computing the full $\mathbf{U}$-matrix.

Starting from an initial solution $\mathbf{p}_0$ and applying Eq. (4.31) to move the solution in the parameter space iteratively, the objective function may be brought down to an acceptable level. The final solution $\mathbf{p}$ contains the fitted values of the parameters such as quadrupole gradients and BPM and corrector gains.

4.2.5   Error analysis

It is important to check the validity of the results when using the least-square method to fit data to a model. A common measure of the goodness-of-fit is the value of the final χ^2. If the fitted model is an exact representation of the experimental system, any deviation between a measured data point and the model prediction can only be due to random measurement errors. The random measurement errors can be assumed to obey Gaussian distributions. If the standard deviations of the error distributions are used to normalize the error terms in the χ^2 definition, the final value of $\chi^2/(N - P)$ after fitting should approach unity, where N is the number of data points and P is the number of fitting parameters.

Substantial deviation of the final $\chi^2/(N - P)$ from unity indicates that the fitted model is a not a true representation of the system. This is the case when the model has systematic errors. Systematic errors can be caused by the omission of some relevant physical processes in the model, or inconsistencies in the data due to drifting experimental conditions during the time span of data taking. It could also mean the fitting algorithm has not converged to the true, global minimum.

In the ideal scenario when the model is accurate and all deviations are random, the uncertainties of the fitted parameters can be derived from measurement errors using error propagation from Eq. (4.30), which relates errors in the fitting parameters, $\Delta\mathbf{p}$, to measurement errors in $\mathbf{r}_0$. It is straightforward to see that the covariance matrix of the fitting parameters is given by

$$\boldsymbol{\Sigma}_{\mathbf{p}} \equiv \langle \Delta\mathbf{p}\Delta\mathbf{p}^T \rangle = (\mathbf{J}^T\mathbf{J})^{-1}\mathbf{J}^T\langle \Delta\mathbf{r}_0\Delta\mathbf{r}_0^T \rangle \mathbf{J}(\mathbf{J}^T\mathbf{J})^{-1} = (\mathbf{J}^T\mathbf{J})^{-1}, \qquad (4.32)$$

where $\langle \cdot \rangle$ denotes ensemble averaging over many possible data sets. It is assumed that $\langle \Delta\mathbf{r}_0\Delta\mathbf{r}_0^T \rangle$ is the identity matrix because $\Delta\mathbf{r}_0$ is normalized by the noise sigma in the measurement and the data points in $\mathbf{r}_0$ are assumed to be uncorrelated. The covariance matrix $\boldsymbol{\Sigma}_{\mathbf{p}}$ contains information of the uncertainties of the fitting parameters. For example, its diagonal elements give the error bars of the fitting parameters,

$$\sigma_{p_i}^2 = [\boldsymbol{\Sigma}_{\mathbf{p}}]_{ii}, \quad i = 1, 2, \cdots, P. \qquad (4.33)$$

As discussed earlier, the Jacobian matrix $\mathbf{J}$ may be degenerate or near degenerate and hence the inversion of $\mathbf{J}^T\mathbf{J}$ can be done only after some small singular values are eliminated. The error sigmas of the fitting parameters calculated this way under-estimate the uncertainties as errors in the directions (in the parameter space) corresponding to the eliminated singular values are not included. In fact, we could calculate the error propagation with Eq. (4.31), which gives

$$\boldsymbol{\Sigma}_{\mathbf{p}} = \sum_{i=1}^{N_{\mathrm{th}}} \frac{1}{s_i^2}\mathbf{v}_i\mathbf{v}_i^T. \qquad (4.34)$$

Therefore, the error bars of the fitting parameters can be given by

$$\sigma_{p_i}^2 = \sum_{j=1}^{N_{\mathrm{th}}} \frac{1}{s_j^2}v_j^2(i), \qquad (4.35)$$

where $v_j(i)$ is the i'th element of vector $\mathbf{v}_j$.

4.2.6 Optics correction and an example

After the least-square fitting, we obtain calibrated values for the fitting parameters, including BPM and corrector gains, BPM and corrector rolls, and quadrupole and skew quadrupole strengths. The BPM and corrector parameters can be used to update the calibration of the corresponding variables in the control system. The lattice model can be updated with the fitted quadrupole and skew quadrupole gradients. The actual lattice parameters that correspond to the conditions of the machine at the time of data taking can be calculated with the calibrated lattice model. The updated lattice model can also be used in tracking simulation to better predict the nonlinear dynamics or collective effects of the real machine.

More importantly, the fitted quadrupole and skew quadrupole parameters give the errors of these parameters relative to their design values, which can be eliminated by adjusting the magnet power supply setpoints. By eliminating the quadrupole errors, the machine linear optics will be corrected toward the ideal condition. Linear optics errors in a storage ring can cause a reduction of the dynamic aperture and the local momentum apertures. Correction of linear optics can lead to improvement in the injection efficiency and the Touschek lifetime. The skew quadrupole errors represent the linear coupling errors and spurious vertical dispersion in the machine lattice. Correction of skew quadrupole errors suppresses the linear coupling and vertical dispersion. In an electron storage ring, this will lead to an reduction of the vertical emittance. In this section we consider only linear optics correction with quadrupole parameters. The BPM and corrector rolls and skew quadrupole parameters are not required in the fitting setup, although they can be included.

In optics correction, the required setpoint change for a quadrupole can be obtained by converting the design gradient, K_0, and the fitted gradient, K_f, to currents using magnet calibration data. The difference of the two currents is the step change required for the quadrupole parameter for optics correction. If the fitting quadrupole parameter corresponds to a power supply that has a non-zero setpoint value, the correction can also be done by scaling the present setpoint by the ratio K_0/K_f.

The application of orbit response matrix fitting for optics correction can be illustrated with a simulated example on the SPEAR3 ring. In the simulation environment [93] all BPM gains are given a random error drawn from a Gaussian distribution with $\sigma = 0.02$. Gradient errors are introduced to three QF magnets and three QD magnets. Orbit response matrix data are taken with the simulator using the method of fixed path length. The corrector kick corresponding to the step change in the measurement is 0.15 mrad for both horizontal and vertical correctors. Dispersion is measured by changing the RF frequency by 1 kHz, corresponding to a momentum deviation of $\Delta\delta = 0.0013$. Both orbit response matrix and dispersion measurements are done in the bipolar mode. Gaussian random BPM errors are added to all measurements. The BPM noise sigma is assumed to be 1 μm for all BPMs.

There are 57 BPMs, 58 horizontal correctors, and 56 vertical correctors. The combined orbit response matrix is 114×114 in dimension. In this test we focus on the linear optics correction; the off-diagonal blocks, $\mathbf{R}_{xy}$ and $\mathbf{R}_{yx}$, are not included in the fitting. The horizontal dispersion is included. Hence the length of the residual vector in the LOCO fitting setup is 6555. The fitting parameters are BPM gains (57×2), corrector gains ($58 + 56$), and 78 quadrupole gradients. The quadrupole gradient parameters include 28 QF magnets, 28 QD magnets, 1 QFC serial power supply, and a number of

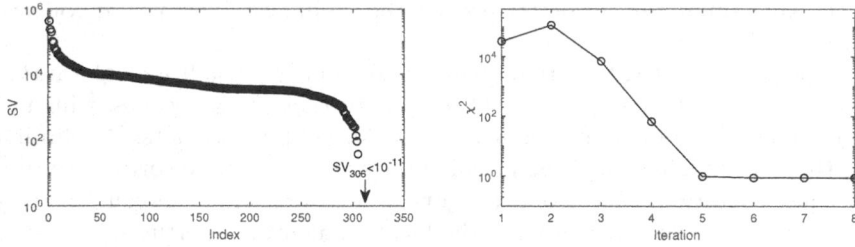

Figure 4.2   Left: singular values of the Jacobian matrix in the LOCO optics fitting for the SPEAR3 setup. The last SV is less than 1×10^{-11} and not shown. Right: the history of $\chi^2/(N - P)$ value over 7 iterations in the test.

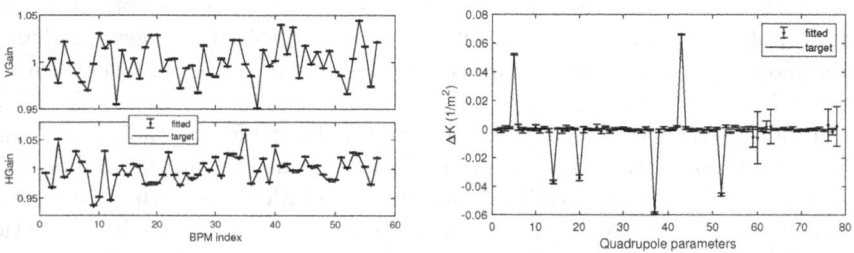

Figure 4.3   Final fitted values in the SPEAR3 LOCO test. Left: BPM gains; right: quadrupole gradient errors, ΔK.

quadrupole magnets in the matching cells and the chicane straight. The total number of fitting parameters is 306.

The singular values of the Jacobian matrix are plotted in Figure 4.2 (left). There is one near-zero singular value, which corresponds to the singularity caused by the simultaneous shifts in BPM and corrector gains in the vertical plane. The inclusion of the horizontal dispersion function has removed the singularity in the horizontal plane. The rest of the singular values range from 40 to 4×10^5. The initial values for all fitting parameters are the ideal values: the BPM and corrector gains are 1.0 and the quadrupole gradients are the design values. The initial normalized χ^2 (per degree of freedom) is 3.27×10^5.

The Gauss-Newton fitting method is applied to the least-square problem (Eq. 4.31), using all but the last singular values. Figure 4.2 (right) shows the normalized χ^2 over 7 iterations. The final value of the normalized χ^2 is 0.89. Figure 4.3 shows the fitted BPM gains and the fitted quadrupole gradient errors ΔK (relative to the design values) in comparison to the target values. The error bars are estimated with error propagation using Eq. (4.35). The fitted BPM gains agree with the gain errors planted in the system when the orbit response matrix data were generated. The fitted quadrupole gradients also successfully recover the errors planted in the lattice. The quadrupole

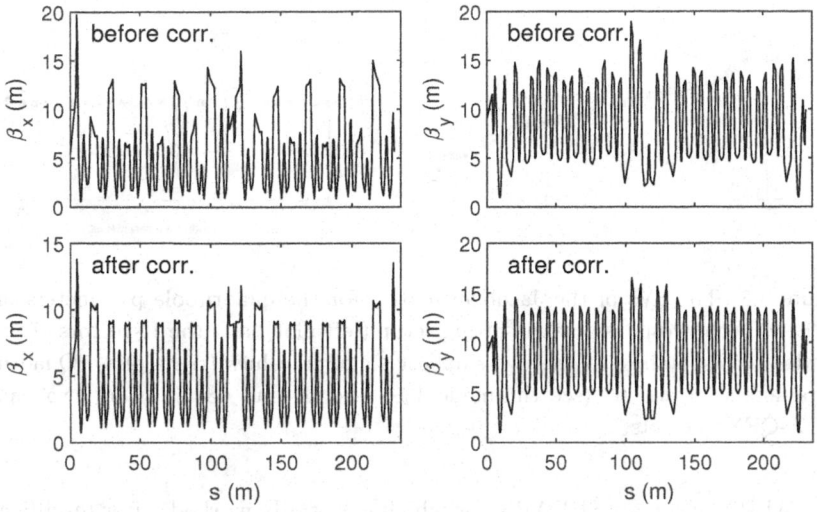

Figure 4.4 Beta functions before (top row) and after optics corrections (bottom row) with LOCO fitting results. The distorted linear optics due to the quadrupole errors is restored toward the periodic optics in the design lattice.

parameters with large error bars correspond to magnets in the chicane long straight section, which are less constrained by the data.

The betatron tunes of the fitted lattice model are $\nu_x = 14.0937$ and $\nu_y = 6.1844$, which agrees with the calculated values for the lattice with the planted quadrupole errors. The beta functions of the lattice with quadrupole errors are severely distorted, as shown in the top row of Figure 4.4. The beta beating relative to the design optics is up to 45% in the horizontal plane and 15% in the vertical plane. After the quadrupole errors are corrected by scaling the setpoint using the fitting results, the betatron tunes are restored to $\nu_x = 14.1065$ and $\nu_y = 6.1753$, very close to the design values of $\nu_x = 14.106$ and $\nu_y = 6.177$. The beta functions also recover the periodicity in the standard cells (bottom row in the figure). The beta beating (in amplitude) of the corrected optics is less than 1.5% in both planes. The residual optics errors in the lattice could be further reduced after a second round of orbit response matrix data taking, fitting, and optics correction.

4.2.7 Constrained least-square fitting

Degeneracy in linear optics fitting problems:

The Gauss-Newton method for orbit response matrix data fitting was successful for many storage rings, including NSLS VUV and X-ray rings [102], ALS, and SPEAR3. However, it failed to work for many other rings, especially the newer storage rings, such as CLS, SOLEIL, DIAMOND, NSLS-II,

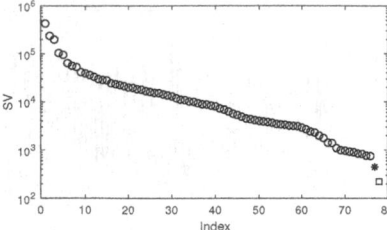 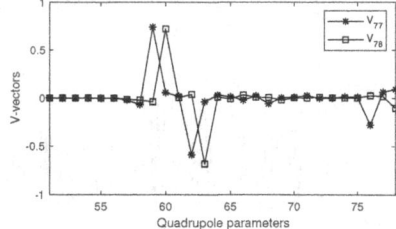

Figure 4.5 The SVs of the Jacobian matrix for the quadrupole parameters in the SPEAR3 orbit response matrix fitting setup (left plot); and the V-vectors of the SV modes corresponding to the two smallest singular values (right plot). Quadrupole parameters 59 and 62 (see the mode V_{77}) and 60 and 63 (V_{78}) are two pairs of QDX-QFX magnets.

and ALBA. Even for SPEAR3, for which it initially worked, after modification to the ring was made to add a chicane in a long straight section, it started having difficulties. The difficulties are due to the similarities in the optics perturbation between the errors of certain quadrupole parameters. When the patterns of optics perturbation by the errors of two quadrupoles are nearly the same, the deviations of the measured orbit response matrix from the design lattice due to the errors will also be nearly the same. Therefore, it is difficult or even impossible to discern the optics error contributions by the two quadrupoles using the orbit response matrix data.

In the SPEAR3 case, when the lattice was changed to accommodate the chicane, the betatron phase advances between the two pairs of quadrupoles next to the chicane straight section were reduced. The two pairs of quadrupole magnets are located at the two ends of the chicane straight, respectively, and each pair consists of a QFX magnet and a QDX magnet. The betatron phase advances between the first pair are $\Delta\psi_x = 0.095$ and $\Delta\psi_y = 0.222$, while the other pair has $\Delta\psi_x = 0.060$ and $\Delta\psi_y = 0.251$. The small differences in phase advances in each pair mean that the two magnets have nearly the same impact to the linear optics and in turn the orbit response matrix. This can be seen from the similarity between the two column vectors in the Jacobian matrix corresponding to the two quadrupole parameters in each pair. The similarity of two column vectors, $\mathbf{v}_1$ and $\mathbf{v}_2$, can be measured by the correlation coefficient,

$$r \equiv \frac{\mathbf{v}_1^T \mathbf{v}_2}{\|\mathbf{v}_1\| \|\mathbf{v}_2\|}, \tag{4.36}$$

where $\|\cdot\|$ is the Euclid norm of the vector. The correlation coefficient for the first pair is $r = 0.9904$, and for the second pair it is $r = 0.9984$.

High correlation coefficients between the columns of a matrix necessarily cause small singular values. In the extreme case, when one column is linearly proportional to another, the correlation coefficient is ± 1 and the matrix becomes rank deficient. If this happens to the Jacobian matrix of the least-square problem, there is a degeneracy involving the corresponding fitting parameters. Figure 4.5 (left) shows the SVs of the Jacobian matrix for the 78 fitting quadrupoles of SPEAR3. The two smallest SVs correspond to the two pairs of correlated quadrupoles. The V-vectors for these two SV modes are shown in the right plot. The last SV mode (V_{78}) primarily represents a near-degeneracy involving quadrupole parameters 60 (QDX) and 63 (QFX) - when these two quadrupoles change in opposite directions, the net impact to the orbit response matrix is small, hence the small SV. Similarly, the second-to-last SV mode (V_{77}) represents another degeneracy the involves quadrupole parameters 59 (QDX) and 62 (QFX). Quadrupole parameters 76-78 are three quadrupole magnets in the chicane straight, which are right between the two pairs of QDX-QFX magnets. There are also correlations between these parameters.

The orbit response matrix data cannot effectively constrain the potential variations of the fitting parameters in the patterns represented by the V-vectors that correspond to small SVs. As shown in Eq. (4.31), a small projection of the residual vector to such an SV mode will generate a large step change to the fitting parameter along the V-vector. Conversely, a large step in the V-vector of the mode only causes a small change to the χ^2 function. SV modes with small SVs may be referred to as under-constrained directions. Consequently, parameters that have large footprints in the under-constrained directions have large error bars, as indicated in Eq. (4.35). In the SPEAR3 case, because of the correlations as discussed in the above, there are large error bars to quadrupole parameters 59, 60, 62, 63, 76, and 78 (see Figure 4.3).

The error bar estimation by Eq. (4.35) only considers the random noise in BPM measurements. Real orbit response matrix data may contain additional errors due to orbit drifts and machine optics fluctuations and other systematic errors. Before the fitting converges, the Jacobian matrix calculated with the model differs from the Jacobian matrix for the measured optics. For example, the differences between the SVs of the Jacobian matrices of the ideal model and that of the actual lattice are up to 60% in the SPEAR3 test case. Furthermore, the expansion of the objective function based on the linearized model of the residual vector (see Eq. (4.29)) may not be accurate when the present solution is far from the minimum. Therefore, during the Gauss-Newton iterations, especially the initial iteration, there could be erroneous projections to the SV modes with small SVs. These projections cause significant changes to the fitting parameters, $\Delta \mathbf{p}$. The resulting solution for the iteration could further deviate from the actual minimum, instead of converging toward it. In some cases, the solution will eventually converge to the minimum. However, if the deviation is large, the fitting method may fail to converge. In an extreme case, the changes to the quadrupole gradients can be so large such that the lattice is not stable (e.g., no closed orbit can be found with the updated model) and hence no further iteration can be carried out [49, 52].

The approach often used in the Gauss-Newton method to deal with small SVs is to set a cut-off threshold and to eliminate all the SVs below the threshold in the calculation of the pseudo-inverse matrix. This approach may work if there are only a few prominent near-degeneracy SV modes, whose SVs are substantially lower than the other SV modes. However, it is common that a large stretch of SV modes all have small SVs. These SV modes can cause large deviations to the fitting parameters. Typically such a solution will have large, unrealistic errors in the quadrupole gradient (e.g., with $\frac{\Delta K}{K}$ at or above a few percent). Because of the large false gradient errors in the solution, when it is applied for optics correction, it could actually increase the optics errors in the machine. On the other hand, removing all the small SVs prevents finding the quadrupole errors in the subspace spanned by the corresponding V-vectors, which in turn will limit the precision of optics correction. Optimizing the cut-off threshold can help balancing the above two detrimental effects; but it does not cure them since either keeping or dropping a small SV can cause problems.

The initial optics correction for the SOLEIL storage ring is a classic example that demonstrates the dilemma of fitting orbit response matrix data with the original LOCO algorithm [88]. There are 56 correctors in each of the two transverse planes and a total of 120 BPMs in the SOLEIL ring. The fitting lattice parameters include 160 quadrupole gradients. When fitting the uncoupled orbit response matrix, there are 512 fitting parameters total. Since there is no clear-cut step on the SV spectrum of the Jacobian matrix, the cut-off threshold had to be found by painstakingly trying out many options. The best option, with 410 SVs kept, predicted $\frac{\Delta K}{K}$ up to 6%, far exceeding the expected gradient errors according to magnet calibration measurements. The beta beating in the machine could not be brought below 5%.

Another approach to combat the near-degeneracy difficulty in orbit response matrix fitting is to reduce the number of fitting quadrupole parameters. By removing some quadrupole parameters that are correlated with other quadrupoles, a set of quadrupole parameters with small correlations between each other can be selected. This approach would work in some cases, but could not address the problem in general. In essence, the approach of removing quadrupole parameters is the same as the approach of removing SV modes: both apply a hard cut to reduce the dimension of the parameter space; the only difference is how to select the dimensions to remove. It could be argued that the approach of removing SV modes is superior because the SV modes are orthogonal and can remove the degeneracy with the least number of dimensions.

Constraints over individual fitting parameters:

In order to use the orbit response matrix data to correct linear optics, a new method is needed to derive a set of reasonable quadrupole errors that can represent the optics errors in the orbit response matrix and are applicable for optics correction. The applicability requires the quadrupole errors to contain only small strays to the SV modes with small singular values. This can be achieved by modifying the Gauss-Newton method with additional constraints to the fitting parameters to prevent large excursions in the under-constrained directions. To introduce constraints on the quadrupole parameters, the least-square objective function is modified to [49, 52, 50]

$$\chi_c^2 = \sum_{ij} \frac{1}{\sigma_{ij}^2}(R_{ij}^{\text{meas}} - R_{ij}^{\text{model}})^2 + \frac{1}{\sigma_K^2}\sum_{i=1}^{N_q} w_i^2 \Delta K_i^2, \qquad (4.37)$$

where σ_K is the uncertainty level of the quadrupole gradients, serving as an overall normalization constant, w_i is the weight factor for the i'th quadrupole parameter, N_q is the number of quadrupole fitting parameters, and ΔK_i is the element in $\Delta \mathbf{p}$ for the i'th quadrupole parameter. For simplicity, the dispersion function terms are not shown in Eq. (4.37). The new terms in the χ^2 definition are called the cost functions. They limit the changes of quadrupole gradients in each iteration. For any change, ΔK_i, to be justified, it needs to cause a considerable reduction to the original χ^2 terms. The cost functions are constraints imposed on the fitting parameters. Changing the weight factors is to change the level of constraints. Because of the constraints, the predicted step change to the fitting parameter vector, $\Delta \mathbf{p}$, will not take large excursions to the under-constrained directions as such excursions tend not to produce significant reduction of χ^2.

With the modified objective function, the residual vector and the Jacobian matrix are changed accordingly. The residual vector is extended to

$$\mathbf{r}_c = \begin{pmatrix} \mathbf{r} \\ \mathbf{r}_w \end{pmatrix}, \qquad \text{with } r_{w,i} = \frac{w_i}{\sigma_K}\Delta K_i, \qquad (4.38)$$

and the Jacobian matrix becomes

$$\mathbf{J}_c = \begin{pmatrix} \mathbf{J} \\ \mathbf{W} \end{pmatrix}, \qquad \text{with } \mathbf{W} = \frac{\partial \mathbf{r}_w}{\partial \mathbf{p}} = \begin{pmatrix} \mathbf{0} & \mathbf{W}_K \end{pmatrix}, \qquad (4.39)$$

where $\mathbf{W}_K$ is a $N_q \times N_q$ diagonal matrix corresponding to the quadrupole gradient fitting parameters, whose diagonal elements are $W_{K,ii} = \frac{w_i}{\sigma_K}$, $i = 1$, $2, \cdots, N_q$, and the $\mathbf{0}$-matrix corresponds to the other fitting parameters. Following Eq. (4.30), the solution of $\Delta \mathbf{p}$ to move from the present solution to the minimum is now

$$\Delta \mathbf{p} = -(\mathbf{J}_c^T \mathbf{J}_c)^{-1}\mathbf{J}_c^T \mathbf{r}_{c0} = -(\mathbf{J}^T \mathbf{J} + \mathbf{W}^T \mathbf{W})^{-1}\mathbf{J}^T \mathbf{r}_0. \qquad (4.40)$$

Compared to the solution by the original Gauss-Newton method (Eq. (4.30)), the constraint terms add a diagonal matrix, $\mathbf{W}^T \mathbf{W}$, to the matrix $\mathbf{J}^T \mathbf{J}$. The addition of the positive definite diagonal terms of $\mathbf{W}^T \mathbf{W}$ (for the quadrupole parameters) to $\mathbf{J}^T \mathbf{J}$ helps eliminate the near-degeneracy - the singular values of the new Jacobian matrix generally move higher.

The weight factors can be determined empirically. A common factor may be given to most of the quadrupole parameters. Elevated weight factors can be assigned to quadrupoles that have high correlations with neighboring quadrupoles. Starting from a low level, all weight factors are then scaled up in steps while observing the ΔK values in the fitting solution and the resulting χ^2. A set of weight factors can be accepted if the fitting solution brings χ^2 down to the same level as the case without constraints (or the case with very weak constraints if no solution can be found without constraints) with reasonable values of $\frac{\Delta K}{K}$. Unless there is a justifiable reason, the fitted $\frac{\Delta K}{K}$ for a quadrupole parameter (relative to the magnet calibration measurement) typically should not exceed the 1% to 2% level. A 1% deviation would be considered high by magnet engineers. However, in practice it is common to see such a level of $\frac{\Delta K}{K}$ in optics fitting results and usually it is applicable for optics correction.

Imposing constraints with Levenberg-Marquardt method:
Adding a positive definite diagonal matrix to the covariance matrix in Eq. (4.40) to improve the convergence is a common practice in nonlinear least-square fitting problems. This method was first proposed by Levenberg and later rediscovered by Marquardt and is referred to as the Levenberg-Marquardt (L-M) method [79, 85]. In nonlinear least-square problems, when the present solution is far away from the minimum, the Gauss-Newton method could fail to converge because the linear expansion of the residual vector may not be sufficiently accurate. In such a case, a sensible approach is to use the gradient-descent method to advance the solution toward the local direction of χ^2 reduction.

The original Levenberg method proposes to modify the Gauss-Newton solution to

$$\Delta \mathbf{p} = -(\mathbf{J}^{\mathbf{T}}\mathbf{J} + \lambda \mathbf{I})^{-1}\mathbf{J}^{\mathbf{T}}\mathbf{r_0}, \tag{4.41}$$

where $\mathbf{I}$ is the identity matrix and λ is a scalar coefficient, which is to be adjusted in each iteration. In an iteration, if the increment $\Delta \mathbf{p}$ does not lead to a reduction of χ^2, the λ value is increased (e.g., by $\times 10$); if it brings down χ^2 by a reasonable amount, λ is reduced. With a large λ, the algorithm behaves like the gradient-descent method since the gradient of χ^2 with respect to $\Delta \mathbf{p}$ is $2\mathbf{J}^T\mathbf{r_0}$ (see Eq. (4.29)). When λ shrinks to near zero, the algorithm becomes the Gauss-Newton method.

A modified version of the L-M method is in the form

$$\Delta \mathbf{p} = -(\mathbf{J}^{\mathbf{T}}\mathbf{J} + \lambda \text{diag}(\mathbf{J}^{\mathbf{T}}\mathbf{J}))^{-1}\mathbf{J}^{\mathbf{T}}\mathbf{r_0}, \tag{4.42}$$

where $\text{diag}(\mathbf{J}^T\mathbf{J})$ is a diagonal matrix that keeps only the diagonal elements of the matrix $\mathbf{J}^T\mathbf{J}$ while setting all other elements to zero. The modified L-M method scales the increment in each axis, such that a larger step can be made in a direction with a smaller gradient.

Comparing the solutions of the L-M method to Eq. (4.40), it is clear that the L-M method is equivalent to imposing constraints to the individual fitting parameters. The corresponding cost functions for the two forms of L-M method are

Original: $\lambda \sum_{i=1}^{N_q} \Delta p_i^2$,

Modified: $\lambda \sum_{i=1}^{N_q} \mathbf{J}_i^T \mathbf{J}_i \Delta p_i^2$, respectively (note that $\mathbf{J}_i^T \mathbf{J}_i$ is a scalar). In the original form, the coefficient λ needs to serve the role of a scaling constant with the dimension of p_i^{-2} for the i'th parameter. Since the parameters may have different dimensions, using a single coefficient for all parameters may not be appropriate. Multiplying each term in the cost function by a scaling constant, e.g, $\frac{1}{\sigma_{p_i}^2}$, would be necessary (with σ_{p_i} being the expected level of uncertainty of parameter p_i). On the other hand, the λ coefficient in the scaled (i.e, modified) L-M method is a dimensionless number independent of the parameters or the least-square problem. Therefore, it is more convenient to use the scaled L-M method.

It is important to note that for the sake of preventing the final solution from taking large excursions to the under-constrained directions, the λ coefficient cannot be allowed to decrease to near zero [50]. If λ is too small, the constraints disappear and the solution may again acquire large ΔK for little reduction of χ^2. Based on the acceptable level of $\frac{\Delta K}{k}$, for any given storage ring, a minimum value of λ may be specified; or simply a constant λ can be used. For the scaled L-M method, $\lambda = 0.001$ is often a good starting point for the initial trial of λ for the orbit response matrix fitting problem.

Constraints over combinations of parameters:

Constraints can also be applied in ways other than adding individual ΔK terms to the χ^2 definition, as was done in Eq. (4.37) or through the L-M methods. For example, if we know a combination of quadrupole parameters are not constrained well by the optics data, a constraint can be imposed to require the stray of the solution in the corresponding direction to be the minimal. A common under-constrained combination is for the gradient of two adjacent quadrupoles to go in opposite directions, if the betatron phase advances between the quadrupole magnets are small. Suppose quadrupole i and $i+1$ are such a pair of neighboring magnets, the term

$$\frac{w^2}{\sigma_K^2}(L_i \Delta K_i - L_j \Delta K_j)^2,$$

can be added to the χ^2 definition to prevent large excursions in the direction represented by $L_i \Delta K_i - L_j \Delta K_j = 0$, where L_i and L_j are the effective lengths of the two magnets, respectively. The residual vector and the Jacobian matrix are then similarly extended, with the difference being that the rows of the Jacobian extension, $\mathbf{W}$, consists of the coefficients that define the direction, e.g., the row for the above constraint term (say, the n'th row of $\mathbf{W}$) would be

$$\mathbf{W}_{n:} = \frac{w_n}{\sigma_K}(0, \cdots, L_i, 0, \cdots, 0, -L_j, 0, \cdots).$$

A general under-constrained direction can be described as

$$\sum_{i=1}^{N_q} c_i \Delta K_i = 0, \qquad \text{or} \quad \mathbf{c}^T \Delta \mathbf{K} = 0, \qquad (4.43)$$

where the coefficients c_i, $i = 1, 2, \cdots, N_q$, define the vector $\mathbf{c}$ in the quadrupole parameter space. It is not easy to intuitively come up with under-constrained directions that are more complex than the simple case with adjacent quadrupoles. However, it is straightforward to determine these directions computationally. In fact, the SV modes of the Jacobian matrix with low singular values are the ultimate representation of the under-constrained directions. To put on constraints to the SV modes, the χ^2 definition can be modified with new terms as follows

$$\chi_c^2 = \chi^2 + \sum_{i=1}^{N_q} \frac{w_i^2}{\sigma_K^2}(\mathbf{v}_{K,i}^T \Delta \mathbf{K})^2, \tag{4.44}$$

where $\mathbf{v}_{K,i}$ are the columns of the $\mathbf{V}$-matrix of the Jacobian matrix (of quadrupole parameters only). The corresponding extended residual vector and Jacobian matrix are in the same form as Eq. (4.38) and Eq. (4.39), with

$$\mathbf{W}_K = \boldsymbol{\Lambda}\mathbf{V}_K, \quad \text{and} \quad \Lambda_{ii} = \frac{w_i}{\sigma_K}, \tag{4.45}$$

where $\boldsymbol{\Lambda}$ is a diagonal matrix with its diagonal elements as given.

In the above we only consider constraints to the quadrupole parameters. This is reasonable as in most cases the difficulties in fitting optics data are caused by the correlation between quadrupole magnets. However, in principle, the constraints can be extended to all fitting parameters. To add constraints on all SV-modes, the least-square objective is modified to

$$\chi_c^2 = \chi^2 + \sum_{i=1}^{P} \lambda_i^2(\mathbf{v}_i^T \Delta \mathbf{p})^2 = \chi^2 + (\mathbf{V}^T \boldsymbol{\Lambda} \Delta \mathbf{p})^2, \tag{4.46}$$

where $\mathbf{V}$ is the $\mathbf{V}$-matrix of the full Jacobian matrix, $\boldsymbol{\Lambda}$ is a diagonal matrix whose diagonal elements, λ_i, $i = 1, 2, \cdots, P$, are the weights on the SV-modes. Correspondingly, the extension to the Jacobian matrix becomes

$$\mathbf{W} = \boldsymbol{\Lambda}\mathbf{V}, \tag{4.47}$$

and, by inserting $\mathbf{J} = \mathbf{U}\mathbf{S}\mathbf{V}^T$ to Eq. (4.40), the solution to $\Delta \mathbf{p}$ for the iteration is now given by

$$\Delta \mathbf{p} = -\mathbf{V}(\mathbf{S}^2 + \boldsymbol{\Lambda}^2)^{-1}\mathbf{S}\mathbf{U}^T\mathbf{r}_0 = -\sum_{i=1}^{P} \frac{s_i}{s_i^2 + \lambda_i^2}\mathbf{v}_i(\mathbf{u}_i^T\mathbf{r}_0). \tag{4.48}$$

Compared to Eq. (4.31), the constraints to the SV modes replace $1/s_i$ with $s_i/(s_i^2 + \lambda_i^2)$ in the calculation of the projected component of $\Delta \mathbf{p}$ in the i'th SV mode. Because of the introduction of the weights, no cut-off threshold of SVs is necessary and all modes can be kept.

It is interesting to note that when the weight factors are the same for all SV modes, i.e., $\lambda_i = \sqrt{\lambda}$ for $i = 1$ through P, Eq. (4.48) is equivalent to Eq. (4.41) (because $\mathbf{I} = \mathbf{V}^T\mathbf{V}$). In other words, when the weights are identical across all terms, adding constraints through individual parameters and through SV modes produces the same results.

4.2.8 Application of constrained fitting for optics correction

As an illustration of the constrained fitting method and also as a demonstration of optics correction with orbit response matrix fitting, in the following we show an example of applying the method to the SPEAR3 storage ring experimentally.

In the test example, the setpoints of a number of QF and QD quadrupole magnets were intentionally changed to introduce optics errors to the machine. Orbit response matrix and dispersion data were taken and fitted with the constrained fitting method. The off-diagonal elements of the orbit response matrix are included in the fitting. In addition to the BPM and corrector gains and 78 quadrupole parameters, BPM coupling coefficients, corrector rolls, and 13 skew quadrupoles are also fitted in order to account for the cross-coupling between the two transverse planes and the vertical dispersion. There are a total of 547 fitting parameters. The constraints are applied to the individual quadrupole and skew quadrupole parameters using the scaled Levenberg-Marquardt scheme.

Figure 4.6 (a) shows the SV spectrum with the weighting factor set to three levels, $\lambda = 0$ (no constraints), 0.001, and 0.1. With the constraints given at a level of $\lambda = 0.001$, the SVs at the low end (e.g., the last 10 SVs) are substantially increased, yet without significant relative changes to the other SV modes. When the weights are increased to $\lambda = 0.1$, however, most of the SVs are significantly increased, leaving only about 15 relatively unchanged. While the modifications made with $\lambda = 0.001$ are essential to prevent excursions to the under-constrained (low-SV) modes, the changes to the SV modes by $\lambda = 0.1$ may have been too large for the solution to quickly converge.

Figure 4.6 (b) shows the convergence history of the normalized χ^2 vs. the Euclid norm of the $\frac{\Delta \mathbf{K}}{K}$ vector for several λ levels (with the additional case of $\lambda = 1 \times 10^{-5}$). The λ value for each case is fixed in all iterations. The Euclid norm of $\frac{\Delta \mathbf{K}}{K}$ serves as an indication of the stray into the under-constrained directions. Without constraints ($\lambda = 0$), the algorithm converges in four iterations, reaching $\chi^2 = 2382$ with $\|\frac{\Delta \mathbf{K}}{K}\| = 0.20$. Its first iteration goes beyond the final solution in terms of the ΔK excursion. The case with $\lambda = 1 \times 10^{-5}$ similarly takes a large first stride in ΔK before returning to a lower level at the second iteration. After it has reached $\chi^2 = 2382$ at the third iteration, it continues on with two more iterations, each resulting in a tiny reduction of χ^2 with a sizable step in ΔK. $\|\frac{\Delta \mathbf{K}}{K}\|$ for the case with $\lambda = 0.001$ does not have an initial overshoot. After χ^2 comes down to 2388 on the third iteration, the fitting algorithm continues to converge with small steps, reaching $\chi^2 = 2383$ on the eighth iteration with $\|\frac{\Delta \mathbf{K}}{K}\| = 0.066$. When the weight is

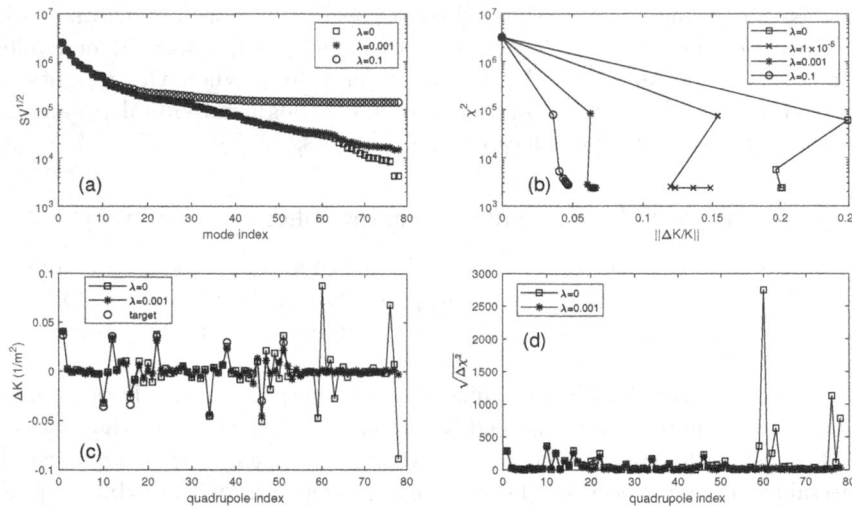

Figure 4.6 Fitting measured orbit response matrix data with the constrained least-square method. (a) The SV spectrum in the first iteration; (b) the history of the normalized χ^2 vs. the norm of $\frac{\Delta K}{K}$ throughout the iterations; (c) fitted quadrupole gradient errors, ΔK, in the final solution for $\lambda = 0$ and 0.001 along with planted errors (9 quadrupoles indicated by circles); (d) $\sqrt{\Delta \chi^2}$, square root of the partial χ^2 contribution for each quadrupole parameter.

increased to $\lambda = 0.1$, the constraints slow down the convergence considerably. After 8 iterations, χ^2 only reaches 2773.

Analysis of χ^2 vs. the level of ΔK excursions such as shown in Figure 4.6 (b) can be used to select the λ value for a given storage ring. In the present example for SPEAR3, it is clear that $\lambda = 0.001$ serves well as it allows quick convergence to the same χ^2 level as the no-constraint case with much smaller ΔK values in the solution. A smaller λ leads to unnecessarily large ΔK excursions, while a large λ causes slow convergence and may distort the solution.

The fitted quadrupole gradient errors for the two cases with $\lambda = 0$ or 0.001 are shown in Figure 4.6 (c). In both cases the quadrupole errors intentionally introduced are found (indicated by circles). But the solution for $\lambda = 0$ has large unexpected errors in quadrupole parameters 59-60, 62-63, and 76-78, which appear in the most severely under-constrained directions. These errors are not found in the solution for $\lambda = 0.001$. Clearly, the latter solution is better suited for optics correction.

The partial χ^2 contribution of a fitting parameter can be used as an indication of the significance of the parameter in the solution. It is defined as the χ^2 variation when the parameter is set to its initial value while all other parameters are at the final fitted values. Figure 4.6 (d) shows $\sqrt{\Delta \chi^2}$ for all quadrupole parameters. Substantial χ^2 changes are seen for the ΔK drifts

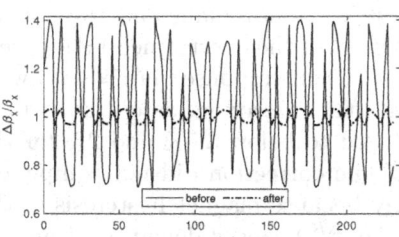 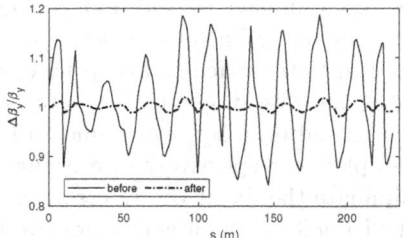

Figure 4.7   Beta beating obtained from fitting orbit response matrix data before and after applying the first round of optics correction to the machine. Left: horizontal; right: vertical.

of the few quadrupoles that dominate the under-constrained directions. However, when all these quadrupoles are restored to the initial values, the total χ^2 change is small, as can be seen from the fact the χ^2 values for the $\lambda = 0.001$ and $\lambda = 0$ cases are practically identical. This reveals that the ΔK drifts are along directions that are inefficient in causing χ^2 changes.

In the present test example, the final χ^2 per degree of freedom is 2382, far greater than 1 as expected in the ideal case. This is because of the systematic errors in the orbit response matrix that are not accounted for by the fitting model. A big part of the residual χ^2 comes from the off-diagonal elements because the 13 skew quadrupoles in the fitting model cannot completely account for the linear coupling in the machine. If the number of skew quadrupoles in the model is increased to 42, the final χ^2 for the data set is reduced to 414. The remaining systematic errors could come from the limited number of lattice parameters that cannot fully account for the actual error sources. Part of it can also come from the inconsistencies in the data as during the time of data taking, the machine conditions (e.g., orbit and optics) could have drifted. Despite the large residual systematic errors, the fitting results are still valid.

For the example in Section 4.2.6, errors in the simulated data set are either from the errors given to the same model parameters being fitted or from the random noise added to the beam orbit. There are no systematic errors that are not accounted for by the fitting model. In this case, χ^2 does converge to the level of 1 as all the residual errors are from the random noise. There is no excessive excursion to the under-constrained directions even though no constraint is applied.

The difference between the fitting results of the simulated example and the experiments (for the no-constraint case, with $\lambda = 0$) suggests that the large excursions in the under-constrained directions in the experiment may be driven by the need to reduce the residual systematic errors, instead of the much smaller share of random errors.

With the fitted lattice model, the linear optics errors can be evaluated. With the quadrupole errors intentionally planted in the machine, the beta beating (amplitude) is as high as 40% in the horizontal plane and 18% for the vertical plane. The quadrupole gradient results of the constrained fitting were applied to the machine for optics correction. Orbit response matrix data were taken and fitted again. The beta beating after correction, as obtained from the second data set, was found to be reduced to below 3.5% and 2% for the two planes, respectively (see Figure 4.7). The correction did not completely eliminate the linear optics errors, probably because magnet hysteresis in the quadrupoles with large changes (up to 3% in $\frac{\Delta K}{K}$) caused deviations from the expected gradient correction. A second round of correction was applied and the beta beating was brought down to below 1% in both planes.

## CHAPTER 5

Linear optics measurement and correction - II

CONTENTS

As discussed in the previous chapter, turn-by-turn BPM data taken with co-herent beam motion sample the lattice similarly as the closed-orbit response matrix. Compared to the latter, turn-by-turn BPM data can be collected within seconds, instead of tens of minutes. There is no complication from calibration errors or rolls of corrector magnets. In addition, linear optics parameters, such as beta functions and phase advances, can be directly measured from the data, before fitting the data to the model. Advanced algorithms can be used to process the data taken by BPMs distributed around the ring in a consistent manner, which allows the separation of random noise, contaminating signals, and unrelated processes from the betatron motion signals.

These methods enable accurate determination of the linear optics. Therefore, turn-by-turn BPM data are ideal for optics measurement and correction.

In this chapter we will discuss various methods of processing turn-by-turn BPM data for linear optics correction. These include the harmonic analysis for betatron phase determination and the 3-BPM method for beta function measurement [19], the model independent analysis (MIA) [65, 119] and independent component analysis (ICA) methods [58], and the method of direct fitting of tracking data [61].

5.1 OPTICS MEASUREMENT WITH HARMONIC ANALYSIS

5.1.1 Phase advance measurement

When the beam undergoes coherent betatron oscillation, the turn-by-turn beam orbit on either the horizontal or the vertical plane seen by a BPM is the discrete sampling of a sinusoidal signal. For example, the turn-by-turn orbit at BPM i is given by

$$x_i(n) = A_i \sin(2\pi\nu n + \psi_i + \chi), \tag{5.1}$$

where n is the turn number, A_i and ψ_i are the oscillation amplitude and the phase advance at the BPM, ν is the betatron tune, and χ is a phase constant common to all BPMs for the motion. The amplitude is related to the beta function at the BPM, β_i, and the action variable, J, via $A_i = \sqrt{2\beta_i J}$. Other frequency components can also show up in the data, for example, the betatron oscillation coupled from the other transverse plane, the synchrotron oscillation through the dispersion function, and potentially signals due to nonlinear coupling resonances. The dominant frequency component is typically the in-plane betatron tune if betatron motion is excited in both planes. Figure 5.1 shows turn-by-turn orbit data on a BPM in the SPEAR3 storage ring during an experiment.

With continuous sampling of the orbit for multiple consecutive turns, the amplitude and phase at the BPM can be determined by calculating the projections of the raw data to the cosine and sine components of the betatron oscillation [19],

$$C_i = \sum_{n=1}^{N} x_i(n) \cos(2\pi\nu n), \quad S_i = \sum_{n=1}^{N} x_i(n) \sin(2\pi\nu n), \tag{5.2}$$

where N is the number of turns. With a large N, the amplitude and phase are given by

$$A_i = \frac{2}{N}\sqrt{C_i^2 + S_i^2}, \tag{5.3}$$

$$\psi_i = \tan^{-1}\frac{C_i}{S_i}, \tag{5.4}$$

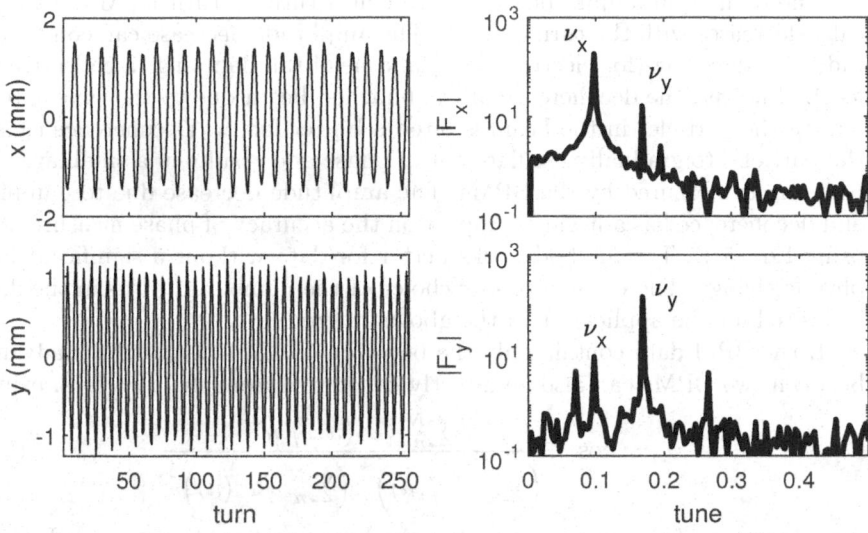

Figure 5.1 Turn-by-turn BPM data taken on SPEAR3 with beam undergoing beta-tron oscillation in both transverse planes. The right plots show the Fourier spectra of the x and y orbit data.

where we have absorbed the phase constant χ in ψ_i. With simultaneous orbit measurements at all BPMs, the beta functions can be determined from the amplitudes A_i, short of a scaling constant related to the action variable. The calibration errors of the BPMs affect the amplitudes. However, the precision of phase measurement is not affected.

The orbit measurements have errors due to random noise in the diagnostics. The errors will affect the accuracy of the phase measurements. Assuming the BPM noise level is σ_x, by the propagation of errors, it can be shown the phase noise sigma is given by

$$\sigma_\psi = \sqrt{\frac{2}{N}} \frac{\sigma_x}{A_i}. \tag{5.5}$$

The accuracy of phase measurements can be improved by increasing the oscillation amplitude and increasing the number of turns.

The use of a finite number of turns in Eq. (5.2) introduces errors to the cosine and sine projections, C_i and S_i, and in turn errors to the phase measurements as the summations in these equations are only approximations to the integrals over a full period. For example, the assumption of $\sum_{n=1}^{N} \sin(2\pi\nu n) \cos(2\pi\nu n) = 0$ is not exact with a finite N for an arbitrary tune. The error due to a finite N can easily exceed the error due to random noise in the orbit and is not mitigated by increasing the oscillation amplitude.

The oscillation amplitude on experimental turn-by-turn BPM data typically decreases with the turn number. The amplitude decrease can come from radiation damping (for electron rings) and head-tail damping (a collective effect). However, the decoherence of the bunched beam due to the tune spread among the particles in the beam is often a bigger factor. Decoherence causes the particles to gradually oscillate out of phase and results in a small average position as measured by the BPMs. The amplitude decrease due to damping and decoherence has a negative impact on the accuracy of phase measurement using Eq. (5.4). The method works better for data without a significant amplitude change. Therefore, a proper choice of the number of turns in the data is desired for the application of the above method.

If the BPM data contain only the betatron oscillation, the phase advance between two BPMs can also be directly calculated from the raw data, using

$$\psi_{ij} = \cos^{-1} \frac{\sum_{n=1}^{N} x_i(n)x_j(n)}{\left(\sum_{n=1}^{N} x_i^2(n)\right)^{\frac{1}{2}} \left(\sum_{n=1}^{N} x_j^2(n)\right)^{\frac{1}{2}}}. \tag{5.6}$$

An advantage of this approach is that the betatron tune is not needed. However, the other frequency components are not separated out and thus will affect the accuracy of the phase measurement. It also suffers from the finite number of turns in the data.

5.1.2 Precise tune determination from turn-by-turn BPM data

To apply Eqs. (5.3)-(5.4), it is important to use an accurate betatron tune in the evaluation of the C_i and S_i coefficients with Eq. (5.2). The tune can be determined from the turn-by-turn orbit with several methods.

Finding peak spectral line: The betatron oscillation recorded on the turn-by-turn BPM data is the same as the discrete sampling of a sinusoidal signal. The frequency spectrum of the data sample can be calculated with the discrete Fourier transform (DFT),

$$F(k) = \sum_{n=1}^{N} x(n)e^{-i2\pi \frac{(k-1)(n-1)}{N}}. \tag{5.7}$$

DFT is usually evaluated with the fast Fourier transform (FFT) algorithm. The coherent betatron oscillation corresponds to a peak at $k = [\nu N + 1]$ on the Fourier spectrum $|F(k)|$, where $[\cdot]$ denotes the closest integer. The betatron tune can be determined up to the accuracy of $\frac{1}{2N}$ by locating the peak spectral line. The precision may not be adequate if there are only tens or a few hundred turns of data.

Fitting in time domain: High precision in the tune determination can be achieved by fitting the turn-by-turn data to a sinusoidal signal as this will accurately locate the tune between two discrete DFT spectral lines. The decaying amplitude due to decoherence and damping can also be included to further improve the accuracy, resulting in the model

$$x(n) = x_0 + Ae^{-\alpha n}\sin(2\pi\nu n + \psi), \tag{5.8}$$

with fitting parameters x_0, A, α, ν, and ψ. The fitting can be done with the least-square method. The initial values of the fitting parameters can be obtained by applying simple analysis to the raw data. For example, the tune can be estimated with FFT and the amplitude and phase by Eqs. (5.3)-(5.4).

The fitting method for tune and phase determination from turn-by-turn orbit data with Eq. (5.8) can yield accurate results and is not sensitive to the finite number of turns. A formula of the phase error due to the finite number of turns is given in Ref. [43]. With the fitting method, the phase is obtained directly as a fitting parameter.

NAFF: The tune can also be determined from the DFT spectrum with other methods that give high accuracy, such as Numerical Analysis of Fundamental Frequency (NAFF) [75] and the interpolated FFT [7]. NAFF is based on the observation that the DFT of discrete samples of a pure sinusoidal signal has spectral lines distributed around the actual tune, which serve as a signature of the tune. The spectrum of a pure sinusoidal signal with the correct tune should be the same as the spectrum of the measured signal in the vicinity of the corresponding spectral line. The similarity can be measured by the projection of the measured signal over a pure sinusoidal signal, $e^{-i2\pi\nu n}$,

$$F(\nu) = \left| \sum_{n=1}^{N} e^{-i2\pi\nu n} x(n) \right|. \tag{5.9}$$

Therefore, the tune can be determined by finding the ν that maximizes the function $F(\nu)$. This is a simple 1-dimensional optimization problem that can be solved using a searching algorithm such as the Nelder-Mead simplex method [89].

Amplitude decay is not considered in Eq. (5.9). This could be included in the model of the pure signal, resulting in a 2-dimensional optimization problem to maximize

$$F(\nu, \alpha) = \frac{\left| \sum_{n=1}^{N} e^{-(i2\pi\nu + \alpha)n} x(n) \right|}{\sqrt{\sum_{n=1}^{N} e^{-2\alpha n}}}, \tag{5.10}$$

where the denominator is the Euclid norm of the reference signal. Essentially, NAFF is to find the tune through data fitting in the frequency domain.

The precision of tune determination with NAFF is $\frac{1}{N^2}$, which is a significant improvement over the simple approach of identifying the peak spectral line. The precision can be further improved by multiplying the raw data by a window function before applying Eq. (5.9) or (5.10). A commonly used window function is the Hanning window

$$W(n) = \sin^2 \frac{\pi n}{N-1}. \tag{5.11}$$

Theoretically, NAFF with Hanning window can achieve the tune precision of $\frac{1}{N^4}$. But that is typically not the case with random noise in the data.

Interpolated FFT: When the tune of the signal does not fall exactly on a spectral line (with interval of $\frac{1}{N}$), the neighboring spectral lines will have nonzero heights. For a pure sinusoidal signal, the heights of the spectral lines can be calculated, which gives [7],

$$|F(\nu_j)| = \left| \frac{\sin N\pi(\nu - \nu_j)}{N \sin \pi(\nu - \nu_j)} \right|, \tag{5.12}$$

where $\nu_j = \frac{j-1}{N}$. Using the heights of the peak spectral line and its highest neighbor, the tune can be determined with the formula

$$\nu = \frac{k-1}{N} + \frac{1}{\pi} \tan^{-1} \frac{|F(\nu_{k+1})| \sin \frac{\pi}{N}}{|F(\nu_k)| + |F(\nu_{k+1})| \cos \frac{\pi}{N}}, \tag{5.13}$$

where k is the peak and $k+1$ its highest neighbor. For a large N, the above formula can be approximated very well by

$$\nu = \frac{k-1}{N} + \frac{1}{N} \cdot \frac{|F(\nu_{k+1})|}{|F(\nu_k)| + |F(\nu_{k+1})|}, \tag{5.14}$$

which is essentially a linear interpolation between the two highest lines.

The precision of tune measurement with interpolated FFT is also $\frac{1}{N^2}$ and it can be improved by data windowing to $\frac{1}{N^4}$ using the Hanning window.

The precision of the tune determination methods can be tested in simulation. In a test, 256 turns of oscillation data with an amplitude of $A = 1$ mm are generated and the tune is determined with 5 methods: fitting in the time domain, NAFF, interpolated FFT (iFFT), NAFF with Hanning window (NAFF+W), and interpolated FFT with Hanning window (iFFT+W). The results are shown in Table 5.1. The top row shows tune errors when no noise is added to the data. The fitting method has the highest precision, followed by NAFF+W and iFFT+W. Random noise with $\sigma = 0.01$ mm is then added to the data. The tune is evaluated for 1000 error seeds. The error sigma of the tune results, σ_ν, is listed in the second row. It is found that data windowing does not improve the precision with noise in the data.

TABLE 5.1 Comparison of the precision of tune determination methods. The methods are applied to 256 turns of data generated with the formula $x(n) = A \sin(2\pi\nu n + \psi)$, and $A = 1$ mm, $\nu = \pi - 3$, and $\psi = \frac{\pi}{6}$. Top row: tune error without noise; Bottom row: the tune error sigma, σ_ν, with noise ($\sigma = 0.01$ mm) added to the data evaluated with 1000 error seeds.

σ/A	Fitting	NAFF	iFFT	NAFF+W	iFFT+W
0	0	2.0E-7	1.1E-5	5.3E-9	3.3E-9
0.01	2.0E-6	1.9E-6	2.3E-6	3.0E-6	3.4E-6

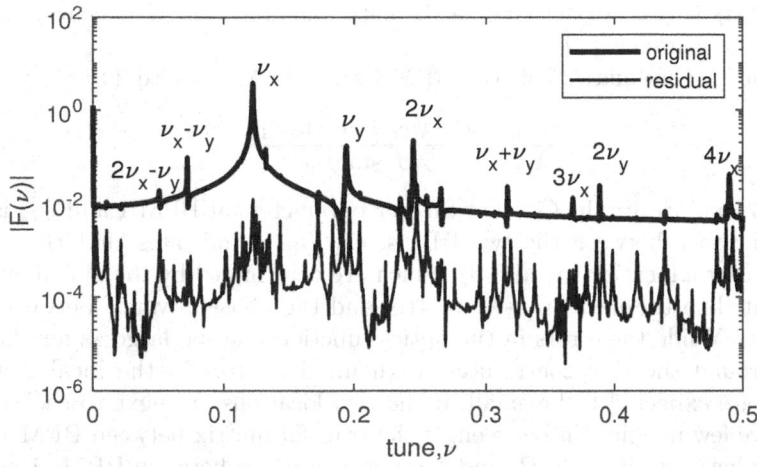

Figure 5.2 Harmonics in simulated horizontal turn-by-turn orbit data are extracted with NAFF. The SPEAR3 lattice is used.

Iterative extraction of harmonics: After the tune is found, the corresponding frequency component in the data can be determined by calculating the projection of the raw data to the component,

$$\tilde{A} = \frac{1}{N} \sum_{n=1}^{N} x(n)e^{-i2\pi\nu(n-1)}. \tag{5.15}$$

The frequency component is then given by $2\text{Re}(\tilde{A}e^{i2\pi\nu(n-1)})$, which can be subtracted from the raw data. The raw data may have many different frequency components. By iteratively applying the procedure of identifying and subtracting the leading frequency component, the harmonics that represent the various resonances in the beam motion can be extracted [74].

As an example, Figure 5.2 shows the Fourier spectrum of the simulated horizontal turn-by-turn BPM data before and after 15 leading harmonics are extracted with NAFF. The oscillation amplitude is 5 mm on both planes. After the major harmonics are removed, the variance of the residual data is substantially reduced.

5.1.3 Determination of beta functions with phase advances

As pointed out earlier, beta function measurement based on the oscillation amplitude at BPMs is not reliable because of the BPM calibration errors. In some cases, it is necessary to determine the beta function to a very high accuracy, for example, for the beta function at the interaction region of a collider.

Given the connection between the beta function and the phase advance (see Eq. (1.75)), it is possible to derive the beta function using the accurate phase advance measurement [19].

If the transfer matrix between BPM 1 and 2 is $\mathbf{A}$, from Eq. (1.68), we have

$$\frac{A_{11}}{A_{12}} = \frac{\cos \psi_{12} + \alpha_1 \sin \psi_{12}}{\beta_1 \sin \psi_{12}}, \tag{5.16}$$

where α_1 and β_1 are the Courant-Snyder parameters at BPM 1 and ψ_{12} is the phase advance between the two BPMs. Eq. (5.16) indicates that the optics functions at a location, α_1 and β_1, which are determined by the global lattice, are related to the local transfer matrix and the phase advance between two locations. While the errors in the optics functions can be large as any lattice error around the ring contributes to them, the errors in the local transfer matrix are expected to be small, if the two locations are next to each other and have few magnets in between. If the transfer matrix between BPM 1 and another location, BPM 3, $\mathbf{B}$, and the phase advance between BPMs 1 and 3, ψ_{13}, are also known, the Courant-Snyder functions at BPM 1 can be solved, using the Eq. (5.16) and its counterpart for BPMs 1 and 3. The beta function is given by

$$\beta_1 = \frac{\cot \psi_{13} - \cot \psi_{12}}{B_{11}/B_{12} - A_{11}/A_{12}}. \tag{5.17}$$

The transfer matrices $\mathbf{A}$ and $\mathbf{B}$ can be calculated with the design lattice model. Using the model lattice functions, Eq. (5.17) can also be rewritten in the form

$$\beta_1^{\text{meas}} = \beta_1^{\text{model}} \frac{\cot \psi_{13}^{\text{meas}} - \cot \psi_{12}^{\text{meas}}}{\cot \psi_{13}^{\text{model}} - \cot \psi_{12}^{\text{model}}}, \tag{5.18}$$

where subscripts "meas" and "model" are used to indicate the measured and model values. An expression for the measured α_1 can be similarly obtained. BPM 3 can be upstream of BPM 1, in which case, ψ_{13} should be interpreted as $-\psi_{31}$ in Eq. (5.18) and the same applies to BPM 2. The method of determining the beta function with phase advance measurement on three BPMs is called the 3-BPM method.

The use of model values of β and ψ in Eq. (5.18) may raise questions about the accuracy of the measured beta function as the model values of these global optics functions differ from the actual values before optics correction. This should not be a concern because the ratio of β_1 and $\cot \psi_{13} - \cot \psi_{12}$ is locally determined. Hence, as long as the differences between the local transfer matrices in the actual machine and the design model are small, measurement of phase advances can be used to deduce the beta function. It is worth noting that for the use of Eq. (5.18), the phase advances ψ_{12} and ψ_{13} should not be close to 0 or $\frac{\pi}{2}$ modulo π because the ratio of β_1 and $\cot \psi_{13} - \cot \psi_{12}$ cannot approach either zero or infinity. The optimal choice is to have phase advances near $\frac{\pi}{4}$ or $\frac{3\pi}{4}$ modulo π.

Three BPMs are used in the above scheme as this is the minimum number of BPMs in order to solve for α_1 and β_1 with Eq. (5.16). To obtain the beta function at one BPM, typically its two immediate neighbors are used if the phase advances satisfy the requirements. Sometimes an immediate neighbor has to be skipped if the phase advance is too close to 0, $\frac{\pi}{2}$, or π.

Using multiple 3-BPM combinations with the two other BPMs going beyond the immediate neighbors can improve the beta measurement accuracy as it brings in a statistical advantage [73]. The measured beta function will be a weighted sum of the selected combinations. The weights are not equal because the phase advance measurement errors of the different combinations are correlated through the common BPMs and also the systematic errors (due to differences in the measured and model transfer matrices) depend on the lattice section between the BPMs. The ideal weights should be derived from the covariance matrix of the measured beta of all 3-BPM combinations, with contributions from both the random phase noise and the systematic errors included. In this case, the method is called the N-BPM method.

The beta functions measured with the 3-BPM or N-BPM methods are not affected by BPM calibrations. By comparing the $\sqrt{\beta}$ measured this way to the betatron oscillation amplitudes, the BPM gains could be calibrated. However, because the beta functions are derived from the phase measurement and thus do not bring in any new information, including these beta functions as input data (in addition to the phase advance measurement) in the linear optics model fitting does not necessarily improve the fitting result.

5.2   GLOBAL ANALYSIS OF TURN-BY-TURN BPM DATA

In the harmonic analysis of the turn-by-turn BPM data, data on different BPMs are analyzed separately. Because of the random noise and the effect of the finite number of turns, the betatron frequency and phase advances obtained from the BPMs may be inconsistent. The results can be substantially improved if all BPM data are analyzed together. This will lead to significantly improved accuracy not only by producing consistent results among all BPMs, but also by taking advantage of the statistics offered by the multiple samples of the same physical processes.

Model independent analysis (MIA) [65, 119] is such a method of global analysis of all BPM data. The statistical analysis method it employs is the principal component analysis (PCA) [97]. In the ideal scenario, through PCA, the various oscillation harmonics in the raw BPM data can be automatically identified and separated. All BPMs share the same time evolution of the harmonics and are thus inherently consistent. Typically there are only a few PCA components that contain information of the beam motion. Since the random noise is distributed over all PCA components, the signal-to-noise ratios in the actual beam motion signals can be improved after the relevant PCA components are isolated. Because the term "model independent analysis" has a very broad meaning and could lead to misunderstanding, in the following we will refer it as the PCA method.

The PCA method can fail to separate the beam motion harmonics when the variances of the harmonics are nearly equal. This could happen when there are bad BPMs, or when there are contamination signals leaked into the BPM electronics. The independent component analysis (ICA) [58] method uses additional features of the underlying beam signals to achieve successful separation despite the above scenarios. Hence it is a much more reliable and robust method to process the BPM data for the global analysis.

In this section we will describe both the PCA and ICA methods and compare the performances.

5.2.1 A model for the turn-by-turn BPM data

BPMs are designed to record the transverse positions of the beam centroid. In a storage ring, ideally, the beam is centered on the closed orbit and there is no orbit change on the turn-by-turn basis. When the beam is displaced from the ideal orbit, the beam will be "excited" and starts to oscillate. The turn-by-turn orbit readings on the BPMs are discrete samples of a continuous motion. Depending on the lattice conditions and the way the beam motion is excited, the beam motion may contain various components, including synchrotron oscillation, betatron oscillations of both transverse planes, and motion from nonlinear resonances. These components are considered source signals. The source signals represent the different physical processes that drive the beam motion. They are considered independent from each other. All BPMs detect the same motion and hence the BPM data contain the same source signals; but the strengths and phases of each component vary with the BPM location.

Considering the orbit signals as linear combinations of the source signals, the raw signal of a BPM can be decomposed as

$$x_i(t) = a_{ij}s_j(t) + \xi_i(t), \qquad (5.19)$$

where subscript i indicates the i'th BPM, $j = 1, 2, \cdots, P$ with P being the number of source signals, and ξ_i is the random noise on the BPM. The orbit oscillation is centered on the closed orbit, which is not of interest here since it is the orbit variation that contains information about the dynamics of the beam motion. For this reason the closed orbit is subtracted from the orbit signal, such that the ensemble average of many turns, $\langle x_i(t) \rangle$, is zero. The a_{ij} coefficients can be scaled up while scaling down $s_j(t)$ by the same factor. The ambiguity can be eliminated by requiring that the row vector $\mathbf{a}_j = (a_{1j}, a_{2j}, \cdots, a_{Pj})$ to have a unit Euclid norm, i.e., $\|\mathbf{a}_j\| = 1$.

The raw signals and source signals can be arranged in vectors, respectively. In a matrix form, Eq. (5.19) becomes

$$\mathbf{x}(t) = \mathbf{A}\mathbf{s}(t) + \boldsymbol{\xi}, \qquad (5.20)$$

where vectors $\mathbf{x}(t)$ and $\boldsymbol{\xi}$ have M elements, one for each BPM, $\mathbf{s}$ has P elements, one for each source signal, and the matrix $\mathbf{A}$ contains all the a_{ij} coefficients. Matrix $\mathbf{A}$ is called the mixing matrix and is $M \times P$ in dimension.

Turn-by-turn BPM data for T turns on all BPMs can be put in a matrix

$$\mathbf{X} = \begin{pmatrix} x_1(1) & x_1(2) & \cdots & x_1(T) \\ x_1(1) & x_1(2) & \cdots & x_1(T) \\ \vdots & \vdots & \ddots & \vdots \\ x_M(1) & x_M(2) & \cdots & x_M(T) \end{pmatrix}, \tag{5.21}$$

where $x_i(n)$ indicates the orbit reading on BPM i for the n'th turn.

Pulse-by-pulse orbit data collected on one-pass systems can be similarly cast into the model of Eq. (5.20). The difference is that the time variation of the source signals does not represent beam motion as each data point is for a different beam pulse. The drifting of the machine conditions with time and the shot to shot jitter will show up on the time variation.

With all turn-by-turn BPM data put in one matrix, the challenge is to separate out the source signals without any additional input.

5.2.2 Principal component analysis (PCA)

Principal component analysis (PCA) is a statistical method for multi-variate data reduction. Suppose the data are samples of M potentially correlated variables, the goal of PCA is to find an orthogonal transformation of the variables to a new set of M variables that are mutually uncorrelated. Furthermore, the new variables are sorted in the descending order in their variances, with each new variable having the maximum possible variance in the remaining subspace after the preceding variables are excluded. In other words, the first new variable is the linear combination of the original variables with the highest variance; the second new variable is the linear combination with the highest variance in the subspace that is orthogonal to the first new variable; and so on.

Mathematically, with the data arranged in a form like Eq. (5.21), PCA is achieved by performing SVD on the data matrix [97],

$$\mathbf{X} = \mathbf{U}\boldsymbol{\Lambda}\mathbf{V}^T, \tag{5.22}$$

with orthogonal matrix $\mathbf{U}$ and $\mathbf{V}$, and diagonal matrix $\boldsymbol{\Lambda} = \text{diag}(\lambda_1, \lambda_2, \cdots, \lambda_M)$ (padded with zeros to make an $M \times T$ matrix). The column vectors in $\mathbf{U}$ represent the distribution of the SVD modes over the BPMs and are called the spatial patterns. The column vectors in $\mathbf{V}$ represent the turn-by-turn variation of the modes and are called the temporal patterns. The orthogonal transformation from the original variables to the new variables is given by

$$\mathbf{Z} = \mathbf{U}^T\mathbf{X}, \tag{5.23}$$

where $\mathbf{Z}$ is the data matrix in the new variables.

The covariance matrix of the raw BPM data is a measure of the distribution of the variance in the multi-variate data space, which is defined as

$$\mathbf{C}_x = \mathbf{X}\mathbf{X}^{\mathbf{T}}, \tag{5.24}$$

noting that each row vector in $\mathbf{X}$ is already zero mean. The covariance matrix of the new variables is a diagonal matrix given by

$$\mathbf{C}_Z = \mathbf{Z}\mathbf{Z}^T = \mathbf{U}^T\mathbf{C}_x\mathbf{U} = \mathbf{\Lambda}\mathbf{\Lambda}^T = \mathrm{diag}(\lambda_1^2, \lambda_2^2, \cdots, \lambda_M^2), \tag{5.25}$$

which shows that λ_i^2, $i = 1, 2, \cdots, M$, are the eigenvalues of the covariance matrix $\mathbf{C}_x$ and the columns vectors in the $\mathbf{U}$ matrix are the corresponding eigenvectors.

The new variables found by PCA are not correlated with one another. This is consistent with the requirement for the underlying source signals. According to Eq. (5.20), the covariance matrix $\mathbf{C}_x$ is related to the covariance matrix of the source signals, $\mathbf{C}_s = \langle \mathbf{s}\mathbf{s}^T \rangle$, via

$$\mathbf{C}_x = \mathbf{A}\mathbf{C}_s\mathbf{A}^T + \mathbf{\Sigma}_n, \tag{5.26}$$

where $\mathbf{\Sigma}_n = \langle \boldsymbol{\xi}\boldsymbol{\xi}^T \rangle = \mathrm{diag}(\sigma_1^2, \sigma_2^2, \cdots, \sigma_M^2)$ is the covariance matrix of the BPM noise. As the source signals are independent of each other, they are necessarily also uncorrelated, and consequently, the covariance matrix $\mathbf{C}_s$ is a diagonal matrix. Eq. (5.26) can be rearranged to give

$$\mathbf{C}_s = \mathbf{A}^{-1}(\mathbf{C}_x - \mathbf{\Sigma}_n)(\mathbf{A}^{-1})^T = \mathrm{diag}(s_1^2, s_2^2, \cdots, s_P^2, 0, \cdots, 0), \tag{5.27}$$

where s_i^2 is the variance of the i'th source signal.

Comparing Eqs. (5.25) and (5.27), we see that if the BPM noise is negligible, i.e., $\mathbf{\Sigma}_n = \mathbf{0}$, we can simply identify the mixing matrix

$$\mathbf{A} = \mathbf{U}, \tag{5.28}$$

and the rows in $\mathbf{Z}$ as samples of the source signals.

Betatron oscillation is typically the dominant component in turn-by-turn BPM data. If we consider only betatron oscillation in one of the transverse planes, the (i, j) element in the BPM data matrix $\mathbf{X}$ is given by

$$x_{ij} = \sqrt{2J\beta_i}\sin(2\pi\nu j + \psi_i), \tag{5.29}$$

where β_i and ψ_i are the beta function and phase advance at BPM i, respectively, and J is the action variable. SVD of $\mathbf{X}$ will find only two non-zero singular values [119],

$$\mathbf{X} = \lambda_1\mathbf{u}_1\mathbf{v}_1^T + \lambda_2\mathbf{u}_2\mathbf{v}_2^T. \tag{5.30}$$

In the case with many turns, the spatial and temporal patterns of the two modes are given by

$$u_{1i} = \frac{\sqrt{JT\beta_i}}{\lambda_1} \cos(\psi_i + \chi), \qquad u_{2i} = \frac{\sqrt{JT\beta_i}}{\lambda_2} \sin(\psi_i + \chi), \qquad (5.31)$$

$$v_{1j} = \sqrt{\frac{2}{T}} \sin(2\pi\nu j + \phi_0), \qquad v_{2j} = \sqrt{\frac{2}{T}} \cos(2\pi\nu j + \phi_0), \qquad (5.32)$$

respectively, where χ and ϕ_0 are phase constants. The phase constant χ can be determined from the orthogonality condition $\mathbf{u}_1 \cdot \mathbf{u}_2 = 0$ and in turn the singular values from the conditions $||\mathbf{u}_1|| = ||\mathbf{u}_2|| = 1$, which give

$$\lambda_1^2, \lambda_2^2 = \frac{JT}{2} \left(\sum_{i=1}^{M} \beta_i \pm \sqrt{(\sum_i \beta_i \cos 2\psi_i)^2 + (\sum_i \beta_i \sin 2\psi_i)^2} \right), \qquad (5.33)$$

where $+$ is for λ_1 and $-$ for λ_2.

From the spatial patterns of the two SV modes, the beta functions and phase advances at the BPMs can be calculated,

$$\beta_i = \frac{1}{JT}(\lambda_1^2 u_{1i}^2 + \lambda_2^2 u_{2i}^2), \quad \psi_i = \tan^{-1}\frac{\lambda_2 u_{2i}}{\lambda_1 u_{1i}}, \qquad (5.34)$$

where we have dropped the phase constant χ.

When random noise is not neglected in Eq. (5.27), the mixing matrix $\mathbf{A}$ would only slightly differ from the eigenmatrix $\mathbf{U}$, provided that the signal to noise ratio is high, i.e., $\sigma_i \ll \lambda_1, \lambda_2$ for all BPMs. Because the BPM noise is uncorrelated with the source signals, it is expected that the noise variance is equally distributed among the new variables (the SV modes). On the other hand, the source signals are concentrated in the few leading modes. Therefore, by reconstructing the data matrix with only the leading modes, the random noise will be reduced. If only the P leading modes are kept, the reconstructed data matrix would be

$$\mathbf{X}_{\mathrm{re}} = \mathbf{U}_P \mathbf{\Lambda}_P \mathbf{V}_P^T, \qquad (5.35)$$

where subscript P indicates only the first P columns in the matrices are kept while the remaining elements are set to zero. The random noise sigma in the reconstructed data will be

$$\sigma_{\mathrm{r}} = \sigma\sqrt{\frac{P}{M}}, \qquad (5.36)$$

where for simplicity we have assumed all BPM noise sigmas are equal to σ.

Figure 5.3 shows an example of the application of PCA to turn-by-turn BPM data. The BPM data are simulated with the SPEAR3 model by tracking 1000 particles randomly generated according to the nominal distribution plus a horizontal position offset of 0.6 mm (w/ $\beta_x = 5.0$ m) and a vertical position

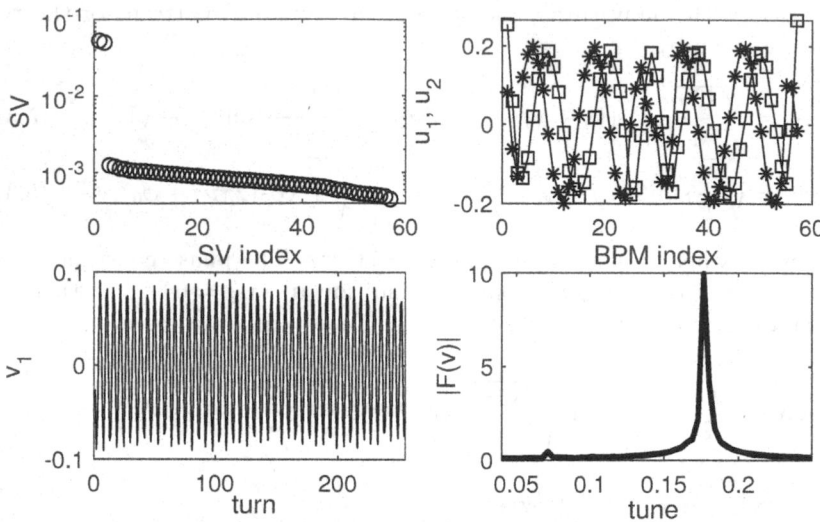

Figure 5.3 PCA for the simulated vertical turn-by-turn BPM data on 57 BPMs over 256 turns for SPEAR3. The SVs (top left), leading spatial patterns (top right), first temporal pattern (bottom left), and the FFT power spectrum of the temporal pattern (bottom right) are shown.

offset of 1.2 mm (w/ $\beta_y = 9.1$ m). The data matrix of the vertical centroid position on 57 BPMs over 256 turns is analyzed with SVD. Random BPM noise is added to all data points, with a noise sigma of 10 μm. There are only two SVs that stand out from the continuous band of small SVs. The small SVs are due to random BPM noise. The two leading SVs correspond to the two orthogonal modes for the betatron motion, which may be referred to as the sine and cosine modes, respectively.

The horizontal and vertical BPM data matrices can be stacked to make a $2M \times T$ overall matrix. With betatron oscillations in both planes, the data matrix will have four non-zero singular values. If the variances in the horizontal and vertical oscillations are not close to equal, the four SV modes consist of two pairs of normal modes - each normal mode represents the motion in a single frequency. In the case of weak linear coupling, the two normal modes can be identified as the horizontal and vertical betatron motions according to their frequencies. The spatial patterns of the horizontal modes contain mostly large amplitudes on horizontal BPMs, and the spatial patterns of the vertical modes mostly on vertical BPMs.

However, if the variances of the two normal modes are about equal, the normal modes may be mixed. This is because if two eigenvalues of the covariance matrix are equal, the corresponding eigenvectors are not uniquely determined. Given the existence of random BPM noise in the data, the normal modes can be more easily mixed in the SVD modes. For example, when the horizontal

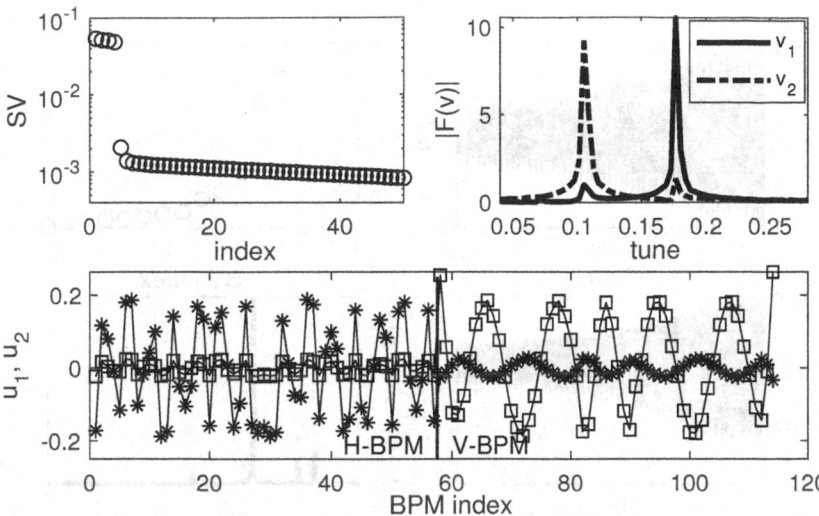

Figure 5.4 PCA for simulated turn-by-turn BPM data in both x and y planes on 57 BPMs over 256 turns. The SV spectrum (top left, partial), the FFT power spectra of the temporal patterns (top right), and the spatial patterns of the first two SV modes are shown. The simulated motion is uncoupled between the two transverse planes, but the horizontal and vertical normal modes are mixed.

and vertical turn-by-turn BPM data of the above simulated case are stacked and analyzed with PCA, the betatron normal modes are mixed in the SV modes, as shown in Figure 5.4. The FFT spectra of the two leading SV modes contain both the horizontal and vertical tunes. The spatial patterns of the two SV modes have substantial amplitudes in the other planes, even though there is no linear coupling in the simulation lattice.

Another common cause of mode mixing in the PCA analysis of turn-by-turn BPM data is bad BPMs. A bad BPM could be very noisy, or have contaminating signals. If the variances of bad BPMs are nearly equal to or larger than those of the source signals, the source signals tend to be mixed with the noise or contaminating signals on the bad BPMs. Figure 5.5 shows the PCA results of an experimental data set taken on SPEAR3 with 8 turn-by-turn BPMs. A contaminating signal with a tune near 0.07 on one of the BPMs causes the failure of separation. A large synchrotron motion component can also cause the failure to separate out the betatron motion. Figure 5.6 shows the results for a horizontal turn-by-turn BPM data set with a longitudinal excitation. The synchrotron motion is not separated from the betatron motion.

When the betatron motion is mixed with other signals, the PCA result is not useful for optics measurements. For a large ring, it could take a considerable effort to identify and remove the bad BPMs before the PCA method can be successfully used.

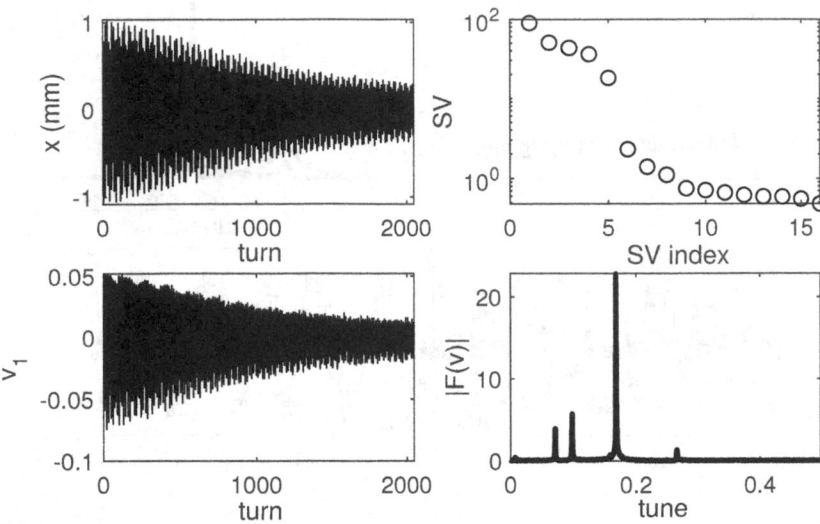

Figure 5.5 PCA results of turn-by-turn BPM data taken on SPEAR3 with 8 BPMs. Clockwise: raw data on one BPM, SV spectrum, temporal pattern and its FFT spectrum of mode 1.

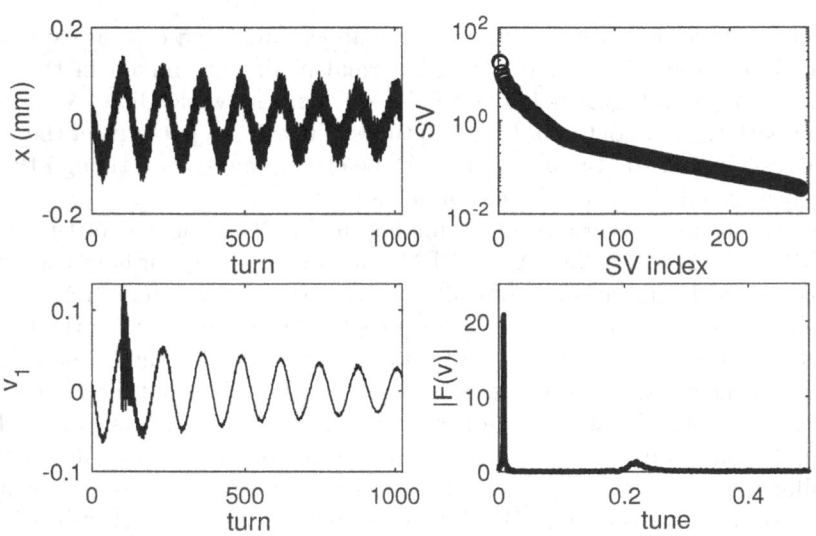

Figure 5.6 PCA results of the horizontal turn-by-turn BPM data measured on a storage ring with an initial longitudinal excitation. Clockwise: raw data on one BPM, SV spectrum, temporal pattern and its FFT spectrum of mode 1.

5.2.3 Independent component analysis (ICA)

By diagonalizing the covariance matrix of the BPM data, PCA finds an orthogonal basis of the multi-variate data space. The new variables defining the orthogonal basis are uncorrelated and would be the underlying source signals if the covariance matrix is not degenerate and there are no noise or bad BPMs. However, in reality there is always noise and it is not uncommon to have bad BPMs. The variances of the source signals can also be nearly equal. Under these circumstances, the uncorrelated variables found by PCA may not necessarily be the source signals. More advanced methods are needed to identify the source signals.

What PCA achieves is the uncorrelatedness in the new variables after a linear transformation. Uncorrelatedness is a requirement for the independent source signals. However, in general, uncorrelatedness does not guarantee independence between the new variables. Two random variables, x_1 and x_2, are uncorrelated if the covariance $\langle x_1 x_2 \rangle = \langle x_1 \rangle \langle x_1 \rangle$. By definition, the two variables are statistically independent when their joint probability distribution is the product of their respective probability distribution functions, i.e., $p(x_1, x_2) = p(x_1)p(x_2)$, from which one can show that for any two functions $h_1(\cdot)$ and $h_2(\cdot)$, $\langle h_1(x_1)h_2(x_2) \rangle = \langle h_1(x_1) \rangle \langle h_2(x_2) \rangle$ is satisfied. Therefore, independence is a much stronger condition than uncorrelatedness. Uncorrelatedness implies independence if and only if the probability distribution functions are Gaussian [64].

The source signals can be identified from the original data by utilizing the additional requirements for the independent variables. This process is referred to as independent component analysis (ICA) [70, 24, 64, 63]. A major category of ICA methods relies on the assumption that the probability distributions of the independent variables are non-Gaussian. These methods try to maximize the non-Gaussianity of the new variables after a linear transformation, using appropriate parametric measures of the non-Gaussianity of the random variables [62].

Other ICA methods exploit the time dependence in the source signals. The time dependence can be fast oscillations or long-term, slow drifts. Separation of source signals from mixed signals recorded on multiple sensors using only the recorded signals without the use of any a priori information is called blind source separation (BSS) [20, 13]. The fast oscillations of the source signals can be characterized by their frequency spectra. ICA methods based on the spectral features are very suitable for the application to turn-by-turn BPM data because the source signals of beam motion in circular accelerators are typically oscillations of certain frequencies. The power spectra of the source signals can be assumed to be narrow-band, for example, a single frequency or with a narrow frequency spread. The mutual independence of two source signals requires their frequency contents to have no overlap – otherwise the two are correlated in the temporal pattern.

The time shifted covariance of two source signals is a way to characterize their spectral correlation. It is defined as

$$C_{12}(\tau) = \langle s_1(t)s_2(t+\tau) \rangle = \int s_1(t)s_2(t+\tau)dt, \qquad (5.37)$$

where τ is the time shift. For continuous signals, we have

$$C_{12}(\tau) = \frac{1}{2\pi} \int F_1(\omega)F_2^*(\omega)e^{i\omega\tau}d\omega, \qquad (5.38)$$

where $F_{1,2}(\omega)$ are the Fourier transforms of $s_{1,2}$, respectively, and * indicates the complex conjugate. Clearly, if the power spectra of the two source signals have no overlap, $C_{12}(\tau) = 0$ for any time shift τ. If $s_2 = s_1$, $C_{12}(\tau)$ becomes the auto-correlation of the source signal, which is equal to the inverse Fourier transform of its power spectrum. Similar results are valid for discrete samples of the source signals. In this case, the time shifted covariance is evaluated by shifting the series of one source signal by τ units and summing up the product $s_1(k)s_2(k+\tau)$.

From Eq. (5.20), we have

$$\mathbf{C}_x(0) = \mathbf{A}\mathbf{C}_s(0)\mathbf{A}^T + \mathbf{\Sigma}_n, \qquad (5.39a)$$

$$\mathbf{C}_x(\tau) = \mathbf{A}\mathbf{C}_s(\tau)\mathbf{A}^T, \quad \text{for } \tau \neq 0, \qquad (5.39b)$$

where the first equation is the ordinary covariance matrix we have seen in Eq. (5.26), the second equation utilizes the fact that the time shifted covariance of the white noise is zero for $\tau \neq 0$. As discussed in the above, the time shifted covariance matrices of the source signals are diagonal matrices. Eq. (5.39b) provides additional information regarding the connection between the data signals and the source signals, which states that the mixing matrix $\mathbf{A}$ not only diagonalizes the usual equal-time covariance matrix, but also diagonalizes all time shifted covariance matrices of the data signals. This extra condition can be used to determine the mixing matrix. Mathematically, the challenge becomes finding a matrix that simultaneously diagonalizes a few time shifted covariance matrices with selected time shifts.

Tools for joint diagonalization of a few matrices have been developed and applied for identification of source signals from mixed data samples. An ICA algorithm, called second order blind identification (SOBI) [13], utilizes this approach and was found to be effective for the application to turn-by-turn BPM data in accelerators. The algorithm consists of two steps. The first step is data whitening, in which PCA is performed to achieve dimension reduction and a normalized, orthogonal representation of the data. The new data matrix becomes

$$\mathbf{z} = \mathbf{\Lambda}_P^{-1}\mathbf{U}_P^T\mathbf{X}, \qquad (5.40)$$

where P SVD modes are kept. The new data matrix satisfies $\langle \mathbf{z}\mathbf{z}^T \rangle = \mathbf{I}_P$, with the P-dimensional identity matrix $\mathbf{I}_P$. The second step is to find an orthogonal matrix $\mathbf{W}$ that joint diagonalizes the time shifted covariance matrix of $\mathbf{z}$ for a few selected time shifts,

$$\mathbf{C}_z(\tau) = \mathbf{W}\mathbf{C}_{\hat{s}}(\tau)\mathbf{W}^T, \quad \text{for } \tau = \tau_1, \tau_2, \cdots, \tau_k, \tag{5.41}$$

where $\mathbf{C}_{\hat{s}}(\tau)$ are diagonal matrices as they are the covariance matrices of the scaled source signals, $\hat{\mathbf{s}} = \mathbf{\Lambda}^{-1}\mathbf{s}$. The choice of the time shifts may depend on the frequency contents of the source signals. For turn-by-turn data, typically no special care is needed. A simple choice such as $\tau = 0, 1, 2, 3$ would work for most cases. Because of noise in the data, joint diagonalization can be achieved only approximately. The numeric algorithm for approximate joint diagonalization is found in Ref. [18]. The source signals and the mixing matrix are then determined by

$$\mathbf{A} = \mathbf{U}_P\mathbf{\Lambda}_P\mathbf{W}\mathbf{\Lambda}_P^{-1}, \tag{5.42}$$

$$\mathbf{s} = \mathbf{\Lambda}_P\mathbf{W}^T\mathbf{z} = \mathbf{\Lambda}_P\mathbf{W}^T\mathbf{\Lambda}_P^{-1}\mathbf{U}_P^T\mathbf{X}. \tag{5.43}$$

When the horizontal and vertical data are processed together in one data matrix, there are two pairs of betatron modes. From the mixing matrix, the beta functions and betatron phase advances can be calculated,

$$\beta_i = a(A_{i,c}^2 + A_{i,s}^2), \quad \psi_i = \tan^{-1}\frac{A_{i,s}}{A_{i,c}}, \tag{5.44}$$

where subscripts c and s indicate the cosine and sine modes for the betatron motion, respectively, i is the BPM index, and a is an overall scaling factor which can be approximately determined by requiring the average beta function over all BPMs to be the same as the model value. There is only one synchrotron mode because typically the phase of synchrotron motion has little change over one turn. The dispersion function can be determined by

$$D_i = bA_{i,d}, \tag{5.45}$$

where subscript d indicates the synchrotron mode and b is a scaling constant.

Figure 5.7 shows two of the ICA modes for the tracking data plotted in Figure 5.4 where PCA fails to separate the horizontal and vertical betatron oscillations due to their equal variances. With ICA, the betatron modes are now completely separated out.

ICA is also applied to the data shown in Figures 5.5 and 5.6. The results are shown in Figure 5.8 and 5.9, respectively. In both cases, the betatron modes are now separated from the contaminating signals or the synchrotron motion.

To evaluate the accuracy of phase advance determination of the ICA method, the betatron phase advances determined from the ICA modes of the simulated data in Figure 5.7 are compared to the values calculated with the lattice model. Betatron phase advances obtained with the harmonic analysis method (using NAFF for tune determination) and the direct sinusoidal fitting

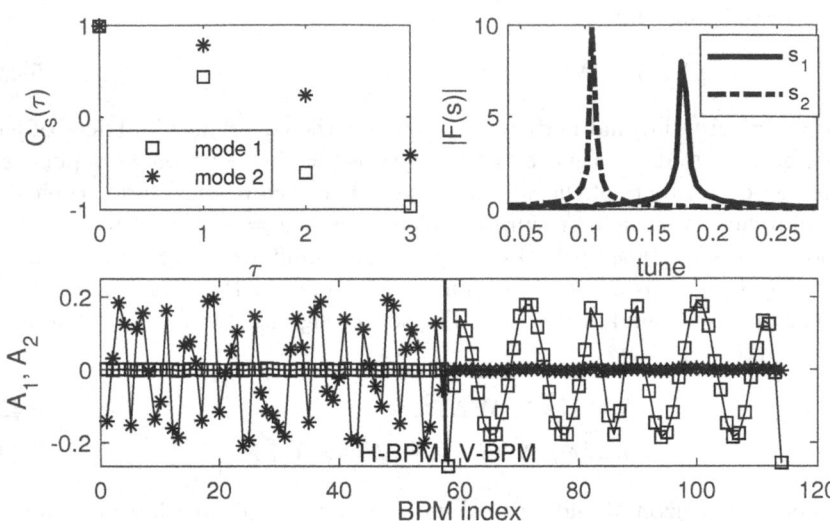

Figure 5.7 ICA is applied to the data in Figure 5.4. The horizontal and vertical betatron modes are completely separated, as can be seen in both the Fourier spectra of the source signals (top right) and the spatial patterns (bottom).

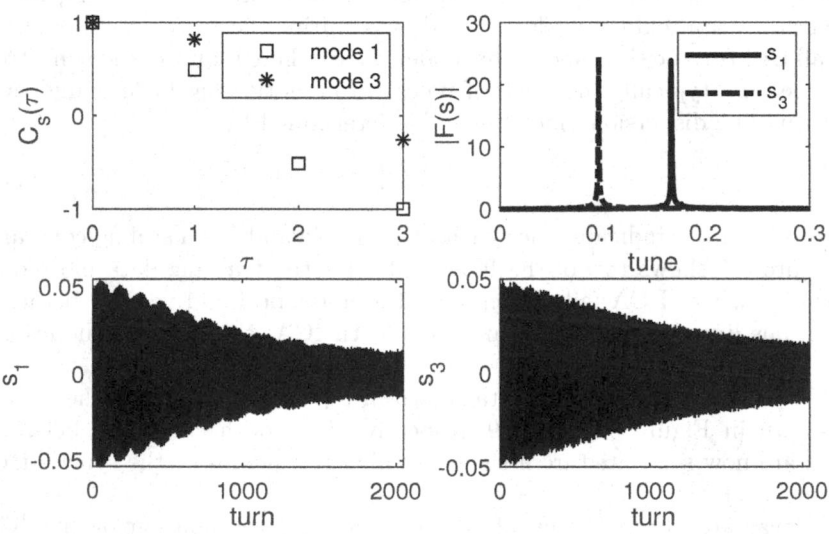

Figure 5.8 ICA is applied to the data in Figure 5.5. The betatron modes are separated from the contaminating signal on one of the BPMs.

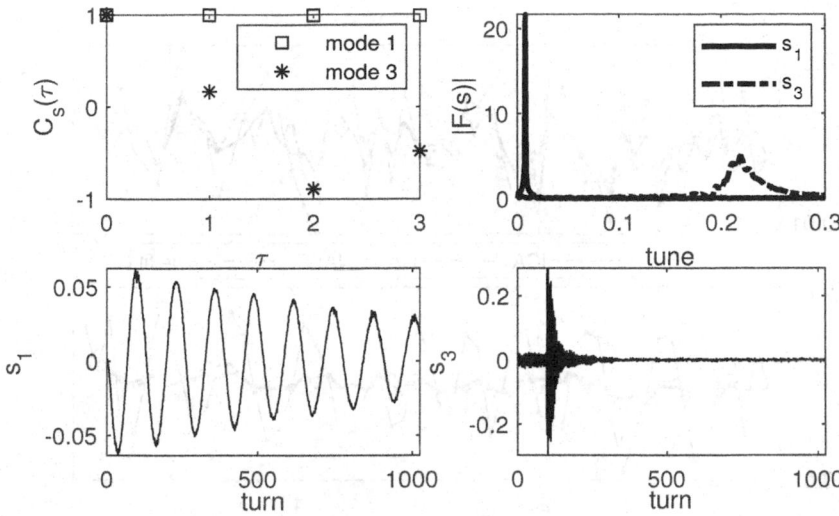

Figure 5.9 ICA is applied to the data in Figure 5.6. The synchrotron motion ("mode 1") and betatron motion ("mode 3") are completely separated out.

are also compared. The results are shown in Figure 5.10 and Table 5.2. Figure 5.10 shows the differences of the measured phase advances from the model values for the three methods for the case with BPM noise sigma of 10 μm. The rms phase errors for the ICA method are 2.5 mrad and 1.7 mrad for the horizontal and vertical planes, respectively. The results by fitting sinusoidal functions are similar. Both are better than the results by the harmonic analysis. The errors in the harmonic analysis results mainly come from the effect of the finite number of turns. As shown in Table 5.2, when the BPM noise is reduced to 1 μm in the simulated data, the errors from the harmonic analysis are about the same as the 10 μm case. However, when the BPM noise is raised to 50 μm, the phase errors are dominated by contributions from the random noise. In this case, the errors for the harmonic analysis and the sinusoidal fit are about the same. The phase errors for the ICA method are lower because of the noise reduction through SVD.

The betatron tune differences between the lattice model and the values derived from the tracking data with NAFF and sinusoidal fitting are shown in Figure 5.11 for all BPMs. The standard deviations of the measured tunes by NAFF are 4.4×10^{-6} and 10.8×10^{-6} for the horizontal and vertical planes, respectively, compared to 2.5×10^{-6} and 2.2×10^{-6} for the fitting method. The tunes derived from the ICA source signals are also plotted. The tune shifts from the lattice model, with $\Delta\nu_x = 5.9 \times 10^{-4}$ and $\Delta\nu_y = 5.0 \times 10^{-4}$, are due to the nonlinear detuning from the finite oscillation amplitude.

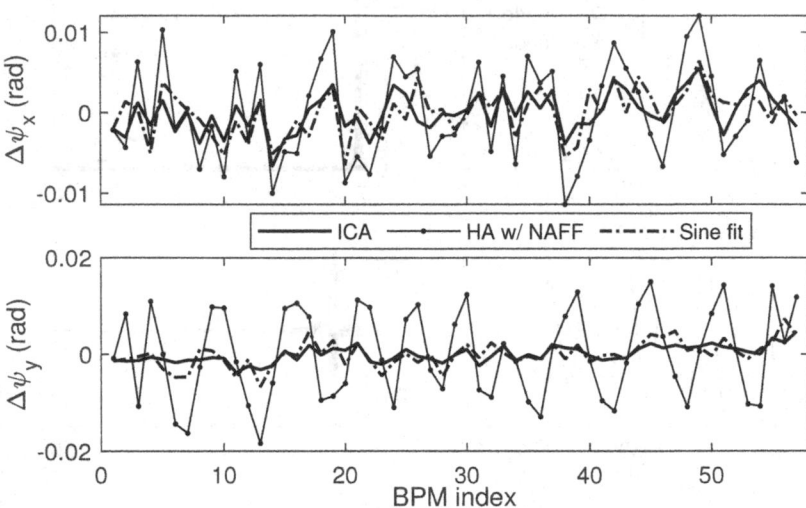

Figure 5.10 Comparison of phase determination accuracy for three methods: ICA, harmonic analysis with NAFF, and the sinusoidal fitting, using simulated data as shown in Figures 5.4 and 5.7. Data consist of 256 turns of turn-by-turn orbits on 57 BPMs with BPM noise sigma of 10 μm. Oscillation amplitude is about 1 mm.

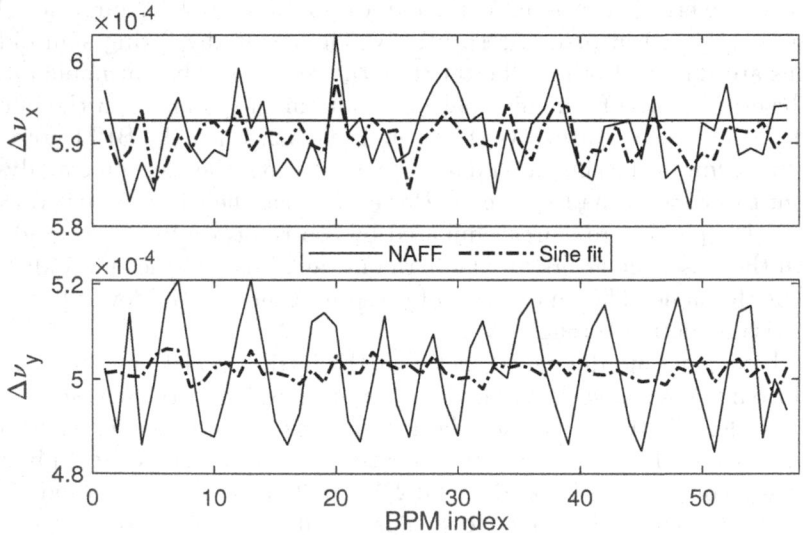

Figure 5.11 Betatron tune differences between the measurement and the model for NAFF, the sinusoidal fitting method, and ICA (straight line).

TABLE 5.2 Comparison of the accuracy of phase advance determination by three methods, with BPM noise sigma set to 1, 10, 50 μm in the tracking data.

σ (μm)	ICA		HA with NAFF		Sine fit	
	$\sigma_{\Delta\psi_x}$	$\sigma_{\Delta\psi_y}$	$\sigma_{\Delta\psi_x}$	$\sigma_{\Delta\psi_y}$	$\sigma_{\Delta\psi_x}$	$\sigma_{\Delta\psi_y}$
1	2.3	1.4	5.6	8.8	1.7	1.0
10	2.5	1.7	5.9	9.3	2.9	2.7
50	5.6	5.2	11.6	14.9	10.4	11.4

5.3 OPTICS CORRECTION WITH TURN-BY-TURN DATA

Measurement of the optics functions is typically not the final goal. The goal is to identify the error sources that cause the optics errors, compensate them, and hence bring the machine optics toward the ideal setting. With turn-by-turn BPM data, two approaches can be used to determine the optics error sources for correction: fitting the measured optics functions to the lattice model, and fitting the turn-by-turn data directly.

5.3.1 Fitting optics functions to lattice model

Beta function and phase advance measurements are direct representation of the linear optics of the machine lattice. The differences between the measured and model values of beta functions and phase advances are optics errors. However, the optics errors do not directly point to the sources of the errors. To correct the optics errors, it is necessary to fit the measured optics functions to the lattice model. This can be done by adjusting the quadrupole gradients in the model to minimize the differences between the measured and model optics functions using the least-square method. This fitting approach is similar to LOCO, the fitting of the orbit response matrix for linear lattice calibration, which was studied in Chapter 4.

In addition to the beta functions and the phase advances, the measured dispersion function can also be included as fitting data. This is necessary if errors in the dispersion function also need to be corrected. The objective function to be minimized is [58]

$$\chi^2 = \sum_i \frac{w_{\beta x}^2}{\sigma_{\beta x,i}^2}(\beta_{x,i}^m - \beta_{x,i}^c)^2 + \frac{w_{\beta y}^2}{\sigma_{\beta y,i}^2}(\beta_{y,i}^m - \beta_{y,i}^c)^2 + \frac{w_{\psi x}^2}{\sigma_{\psi x,i}^2}(\psi_{x,i}^m - \psi_{x,i}^c)^2 +$$

$$\frac{w_{\psi y}^2}{\sigma_{\psi y,i}^2}(\psi_{y,i}^m - \psi_{y,i}^c)^2 + \frac{w_{Dx}^2}{\sigma_{Dx,i}^2}(D_{x,i}^m - D_{x,i}^c)^2, \tag{5.46}$$

where superscripts 'm' and 'c' indicate the measured and calculated values, respectively, the σ's are the error sigmas of the measured values, and the w's are the weights assigned to the data types. The error sigmas can be estimated by evaluating the standard deviation of the results from multiple data sets.

Error propagation may also be used, for example, for beta functions and phase advances obtained through fitting the sinusoidal model. The weights are included to adjust the relative importance of the data types. This is necessary because there could be systematic errors in the data types. For example, beta functions obtained with the sinusoidal fitting method and the ICA method are subject to BPM calibration errors. In some cases, the beta functions can be excluded by setting $w_{\beta x} = w_{\beta y} = 0$. This is acceptable because the errors in the beta functions and the phase advances are closely related (see Eqs. (2.27) and (2.30)) – to some extent the beta beating data could be seen as redundant information. The weight of the dispersion function can be increased to achieve a high precision of dispersion control.

The fitting parameters are the quadrupole gradients. These can usually be the same quadrupole parameters as the orbit response matrix fitting. The terms in χ^2 can be represented by the residual vector, $\mathbf{r}$, such that the objective function takes the standard form, $\chi^2 = f(\mathbf{p}) = \mathbf{r}^T \mathbf{r}$. The least-square fitting methods discussed in the previous chapter can then be applied. The degeneracy problem due to the cross-coupling between adjacent quadrupole magnets is common to all optics fitting, including the present case with optics functions as input data. Therefore, the constrained fitting technique in section 4.2.7 is generally needed.

Optics correction with ICA for turn-by-turn BPM data analysis has been experimentally demonstrated on the RHIC collider [109] and on the NSLS-II storage ring [51, 125]. In the NSLS-II experiment 1024 turns of orbit data from 180 BPMs are decomposed into two pairs of betatron normal modes and the synchrotron mode. The Fourier spectra of the data from one BPM (located at a dispersive region) and the temporal patterns of the normal modes are shown in Figure 5.12 (top plots). The amplitudes of the normal modes on the horizontal (middle plot) and vertical (bottom plot) BPMs are also shown. The oscillation amplitude is about 0.2 mm. Because of linear coupling, the vertical betatron oscillation shows up on the horizontal orbit and vice versa.

From the spatial patterns, the betatron phase advances of the normal modes are calculated. The differences of the measured phase advances with the ideal model are shown in Figure 5.13. The values measured by ICA are compared to the lattice model calibrated with the orbit response matrix data taken at the same time as the turn-by-turn BPM data. The differences between the measurements by the two methods are small. The beta functions, phase advances, and the dispersion function are used to fit the lattice model using Eq. (5.46). The fitted BPM gains and quadrupole gradient errors are compared to the orbit response matrix fitting results in Figure 5.14 and 5.15, respectively. The fitting results between the two methods are very similar. Because of the different constraints applied, the fitted $\frac{\Delta K}{K}$ by ICA is slightly smaller than LOCO, while the two fitted lattices have almost the same optics errors.

The fitted quadrupole errors by ICA were applied to the machine for optics correction. After three iterations of corrections, each time using a new data set of turn-by-turn BPM data taken on the updated machine, the optics errors

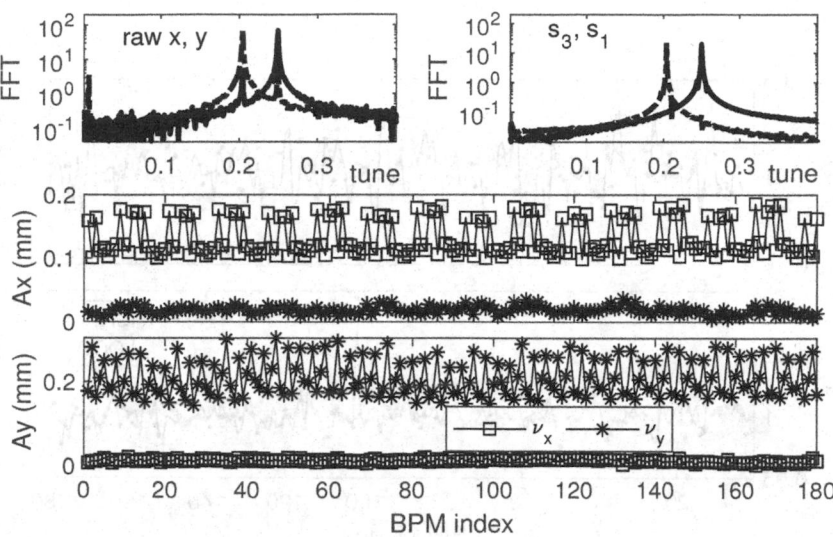

Figure 5.12 The betatron modes of the turn-by-turn BPM data taken on NSLS-II. Plotted are the Fourier spectra of the raw x and y data on BPM 3 (top left) and the two ICA normal modes (top right), the betatron amplitudes of the two normal modes on the horizontal BPMs (middle) and the vertical BPMs (bottom).

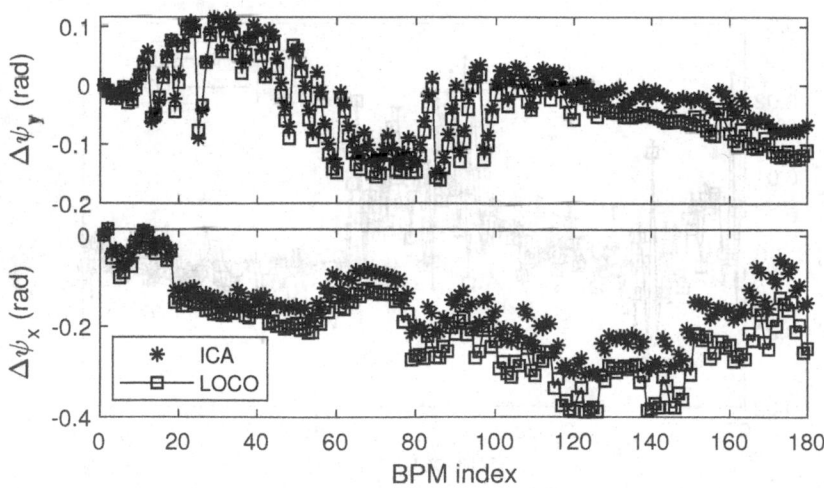

Figure 5.13 Differences between the measured phase advances and the ideal model for NSLS-II measured from the turn-by-turn BPM data with ICA ("ICA") or from fitting the orbit response matrix ("LOCO").

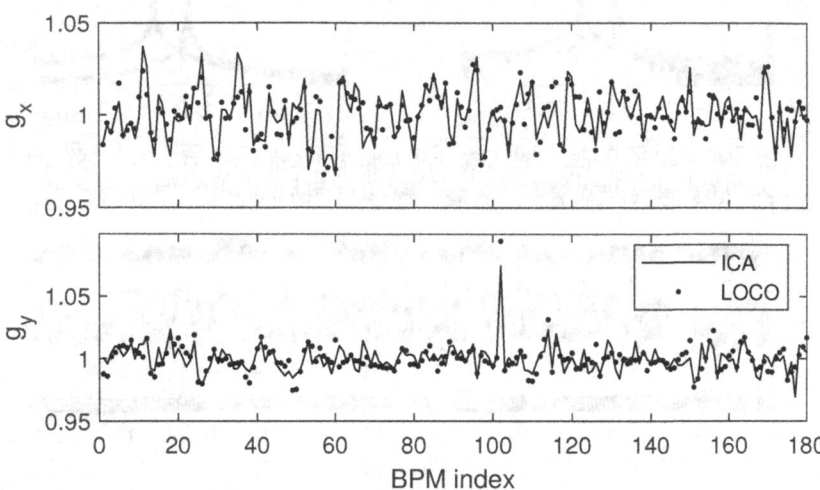

Figure 5.14 The fitted horizontal (top) and vertical (bottom) BPM gains for NSLS-II by ICA ("ICA") or by fitting the orbit response matrix ("LOCO").

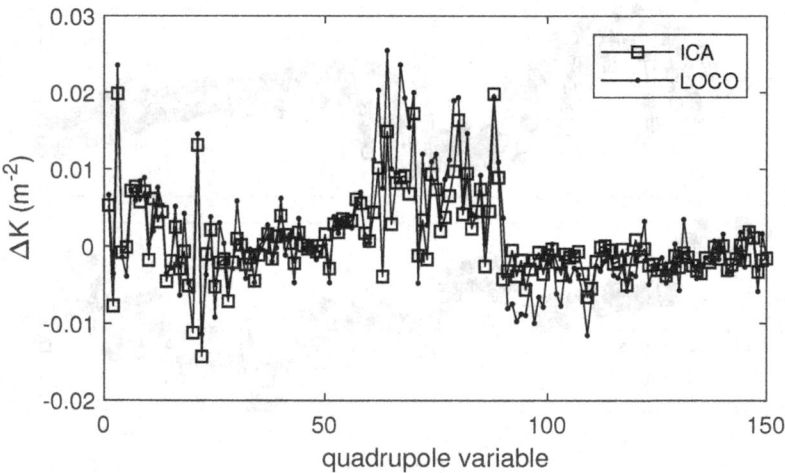

Figure 5.15 The fitted quadrupole gradient errors, ΔK, for NSLS-II by ICA or by fitting the orbit response matrix ("LOCO").

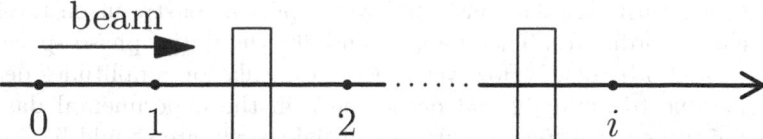

Figure 5.16 Fitting the lattice model directly with BPM data: deriving the angle coordinates with BPMs 0 and 1, predicting beam positions with particle tracking, and comparing the measured and predicted positions.

were substantially reduced. The rms beta beating decreased from 7% (H) and 9% (V) to 0.5% (H) and 0.4% (V), respectively, while the rms phase advance beating decreased from 30 mrad (H) and 40 mrad (V) to 8 mrad (H) and 7 mrad (V). The rms horizontal dispersion errors also decreased from 17 mm to 5 mm. The precision of optics correction could have been better if the signal-to-noise ratio was higher in the turn-by-turn BPM data (oscillation amplitude was only 0.2 ∼ 0.3 mm).

5.3.2 Fitting turn-by-turn data directly to lattice model

Turn-by-turn BPM data can also be directly fitted to the lattice model if the data can be predicted with particle tracking simulation. Full phase space coordinates are needed for tracking simulation. The BPMs only measure the position coordinates. However, the angle coordinates can be calculated from the position coordinates at two BPMs with a known transfer matrix in between, as shown in Eqs. (4.5) and (4.6). With the full transverse phase space coordinates at one BPM, $\mathbf{X}=(x, x', y, y')^T$, the coordinates at all BPMs over subsequent turns can be obtained from tracking and compared to the measured data. The BPM layout in the lattice is as illustrated in Figure 5.16, where BPMs 0 and 1 are separated by a drift space. To avoid large errors to the angle coordinates, it is desirable to have a large distance between the two BPMs. Since we are trying to extract the optics information from the coherent beam motion, the closed orbit is subtracted from the BPM readings.

Lattice parameters can be adjusted to minimize the difference between the measurement and the tracking data, which is characterized by the least-square objective function

$$\chi^2 = f(\mathbf{p}) = \sum_{i=1,k=1}^{M,T} \frac{(x_i^{\mathrm{m}}(k) - x_i^{\mathrm{c}}(k;\mathbf{p}))^2}{\sigma_{xi}^2} + \frac{(y_i^{\mathrm{m}}(k) - y_i^{\mathrm{c}}(k;\mathbf{p}))^2}{\sigma_{yi}^2}, \qquad (5.47)$$

where i and k are the BPM and turn indices, respectively, σ_{xi} and σ_{yi} are the horizontal and vertical BPM noise levels, x^{m} and y^{m} are measured beam positions, and x^{c} and y^{c} are positions obtained with tracking. The phase space coordinates can be tracked for multiple turns and used for comparison, hence the summation of turns in Eq. (5.47). The measured orbits at multiple BPMs over multiple turns are self consistent, which puts a constraint on the initial phase space coordinates. Therefore, potentially, the initial phase space coordinates can also be fitted. However, since the oscillation amplitudes decrease with time due to damping and decoherence in the experimental data, the number of turns to be tracked with one initial coordinate should be limited.

The fitting parameters can include the quadrupole gradients and the BPM gains when only the linear optics is concerned. Skew quadrupole gradients and BPM rolls and crunch coefficients can also be fitted to account for the cross-plane coupling.

The direct fitting scheme can also be applied to one-pass lattices. In this case there is no summation over turns and the summation over BPMs starts with $i = 2$ in Eq. (5.47). To adequately sample the phase space, a pair of corrector magnets upstream of BPMs 0 and 1 for each transverse planes are required to scan the beam trajectory. In each plane, it is necessary to scan many combinations of (x, x') or (y, y') with phase angles that cover the full range of $[0, 2\pi)$. Different phase angles are needed because the data points with the same phase angle, as obtained by scaling the strengths of both correctors in proportions, are redundant information. Lattice fitting for rings can be treated in the same manner as for the one-pass systems by considering the ring as a transport line, using BPM data of each turn as an independent sample.

Because of the lack of temporal betatron oscillations, the methods of deriving phase advances and beta functions described earlier in this chapter do not apply to one-pass lattices. The direct fitting scheme is particularly important for transport lines and linacs. It can also be applied to the commissioning of storage rings before the stored beam is established. The BPM data fitting method has been experimentally demonstrated using a section of the SPEAR3 storage ring [61] and the LCLS linac and transport lines [128].

In the LCLS linear optics fitting study [128], both horizontal and vertical trajectories are scanned on a phase space grid with two pairs of corrector magnets upstream of the lattice of concern. For the linac section, backward tracking was used as the pair of BPMs separated by a long drift are located downstream of the linac. As the beam energy increases in the linac, the action variables for (x, x') or (y, y') decrease. This effect is accounted for in the tracking model. The fitted quadrupole gradients successfully recovered the strengths of the matching quadrupoles. There were substantial differences between the fitted quadrupole gradients and the ideal model for a few magnets as they were manually tuned during operation. For the transport lines after the linac, both forward and backward tracking were used and the results were consistent with the operation setting.

The direct fitting method for storage ring optics measurement can be tested with the NSLS-II turn-by-turn BPM data discussed in the previous section. The pair of BPMs separated by the first long straight section are used to calculate the initial phase space coordinates. The two BPMs are separated by a distance of 9.87 m and there are two harmonic sextupoles (SH1) in between, next to the BPMs. The nonlinear kick by the second sextupole is not

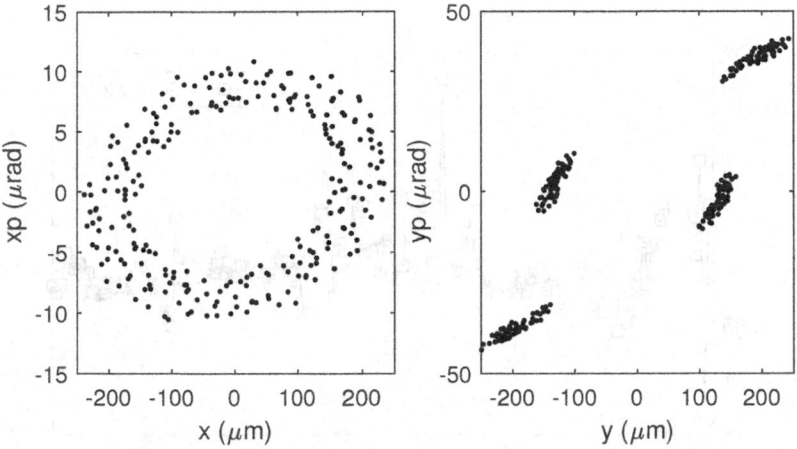

Figure 5.17 The initial phase space coordinates on BPM 1 derived from a pair of BPMs for the NSLS-II lattice fitting.

included in the calculated angle coordinate. It can be corrected with the magnet strength and the position coordinates. The horizontal and vertical kicks are up to 0.1 μrad and 0.2 μrad, respectively, compared to the maximum angle coordinates of 10.8 μrad (H) and 42.3 μrad (V). In one test, 256 turns of orbit data are used and the coordinates are tracked for one turn for comparison. In this setup the ring is treated almost as a transport line, except the next-turn positions on BPM 0 and 1 are also compared. The initial 256 turns of phase space coordinates on BPM 1 are as shown in Figure 5.17. The BPM noise sigma is estimated to be 2.6 μm, using the last 350 SVD modes (out of 360 total modes). Noise reduction with SVD could improve the fitting result, but was not used in the test.

Fitting with the Levenberg-Marquardt method with an initial $\lambda = 0.01$ for two iterations reduced the normalized χ^2 from 62 to 5.1. The fitted BPM gains differ from the ICA fitting results (see Figure 5.14) by only 0.009 (H) and 0.007 (V) (rms values) for the two planes, respectively. The fitted quadrupole gradient errors are shown in Figure 5.18. The results are very similar to the ICA fitting results. The phase advance errors of the fitted lattice are shown in Figure 5.19. The phase advance differences between the lattices fitted with the ICA results and the direct BPM data are 4.6 mrad (H) and 4.0 mrad (V) (rms), respectively. The rms beta beatings between the two fitted lattices are only 1.5% (H) and 0.7% (V), respectively. Optics correction with the direct position fitting would have achieved the same level of beta beating reduction as fitting the ICA results.

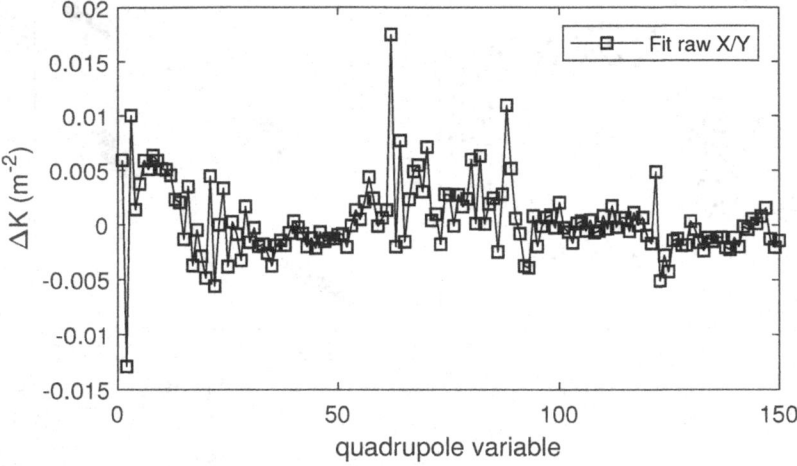

Figure 5.18 The fitted quadrupole gradient errors, ΔK, for NSLS-II by directly fitting turn-by-turn position data, using the least-square method for Eq. (5.47).

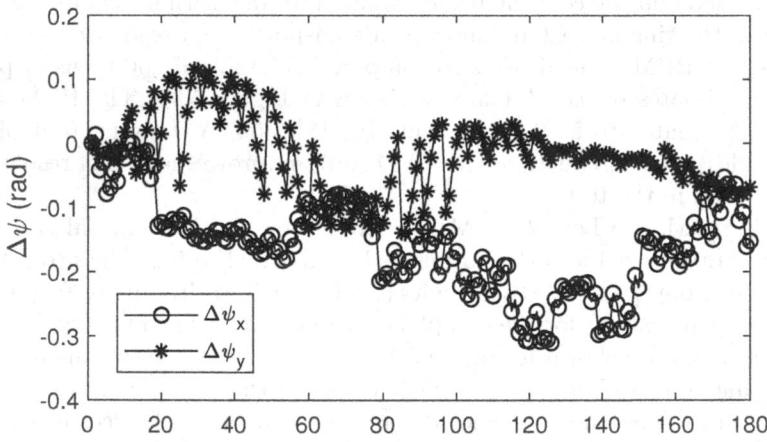

Figure 5.19 The phase advance errors in the fitted lattice obtained by fitting the raw position data, to be compared to Figure 5.13.

5.4 OTHER METHODS FOR OPTICS CORRECTION

Other methods can also be used to extract the linear optics errors from the turn-by-turn BPM data for lattice model calibration and optics correction.

Segment-by-segment optics function fitting: With the 3-BPM or N-BPM method, the Courant-Snyder parameters can be determined from the simultaneous turn-by-turn BPM data at the BPM locations. For a large ring, the lattice can be divided into several segments and treated separately. Using the α and β functions at the beginning of the segment, the Courant-Snyder parameters and phase advances at downstream BPM locations can be calculated with the lattice model. The differences of these optics functions between the measurements and the calculation can be used to fit the quadrupole gradient errors in the lattice segment, from which the model can be calibrated and the results can be used for optics correction. This method was used for the LHC optics measurement and correction [3, 116] and has also been extended to electron storage rings [73, 84].

Resonance driving terms (RDTs): In general, turn-by-turn beam motion can be decomposed into a series of harmonics of the betatron tunes, as is shown in Eqs. (2.116)-(2.117). If the beam motion is purely linear and the linear optics is the same as the design model, the turn-by-turn motion in the resonance basis coordinate (using the design optics functions to obtain the normalized betatron coordinates) is a simple rotation. However, with linear optics errors, the motion of the resonance basis coordinates will be distorted.

The linear optics errors in the horizontal and vertical planes are characterized by the f_{2000} and f_{0020} terms, respectively, which are related to the quadrupole errors distributed throughout the ring via [10, 39]

$$f_{2000} = \frac{\sum_k \Delta b_{1,k} L_k \beta_{x,k} e^{i2\psi_{x,k}}}{8(1 - e^{i2\pi\nu_x})}, \quad f_{0020} = -\frac{\sum_k \Delta b_{1,k} L_k \beta_{y,k} e^{i2\psi_{y,k}}}{8(1 - e^{i2\pi\nu_y})}, \quad (5.48)$$

where $\Delta b_{1,k}$ and L_k are the normalized gradient errors and the lengths of the quadrupole error sources, respectively, and $\beta_{xy,k}$ and $\psi_{xy,k}$ are the beta functions and the phase advances (between the error sources and the observation point), respectively. The corresponding spectral lines are $1-\nu_x$ on h_x^- for f_{2000} and $1 - \nu_y$ on h_y^- for f_{0020}, respectively.

By using two adjacent BPMs to determine the angle coordinates, the turn-by-turn resonance basis coordinates $h_{x,y}^-$ can be calculated (see Eq. (2.109)). The amplitude and phase of f_{2000} are subsequently determined from the component corresponding to the $1 - \nu_x$ tune line on the Fourier spectrum of h_x^- (see Eq. (2.116)) using NAFF or interpolated FFT. Similarly f_{0020} is determined from the $1 - \nu_y$ tune line on h_y^-. If this can be done at many locations around the ring, the RDT data can be used to fit the quadrupole errors in the lattice with Eq. (5.48) [39, 47].

Transfer matrices: When the angle coordinates are calculated with position data from two adjacent BPMs, the phase space coordinates can be used to determine the transfer matrices, using Eq. (4.8). From the one-turn transfer matrix the Courant-Snyder parameters can be calculated. Using the transfer matrices between two locations, the phase advances can also be determined. These derived data can then be fitted to the lattice model to determine the quadrupole errors. One may also fit the lattice by directly comparing the transfer matrices between the model and the measurements.

5.5 SUMMARY FOR LINEAR OPTICS CORRECTION

Linear optics correction for storage rings has been an extensively researched area. LOCO, the method of fitting the orbit response matrix [102], was an early success, which became the standard method for optics correction on electron storage rings. The constrained fitting technique was an important development [49, 50] for the method as it solved the degeneracy problem that caused the original fitting method to fail for many machines and limited the precision of optics correction. The improved fitting method is the default method in the LOCO code [95] in the form of the scaled Levenberg-Marquardt method.

One limitation of the LOCO method is that the measurement of the orbit response matrix is time consuming. For the SPEAR3 ring it takes 10 minutes to measure one data set. For large rings or rings with slow ramping correctors, the time would be considerably longer. A new approach that uses fast correctors to modulate the beam orbit at different frequencies can substantially reduce the data taking time (e.g., from 1 hour reduced to 2 minutes for NSLS-II) as the responses of multiple correctors can be simultaneously measured [124].

In recent years, as turn-by-turn BPMs are widely used on new rings, methods based on simultaneous turn-by-turn BPM data have been developed and studied at many facilities [58, 3, 61, 116, 109, 125, 47]. Most methods rely on the accurate phase measurements from the turn-by-turn BPM data, which are used to fit the lattice model. There have been a number of studies that compare the performances of LOCO and turn-by-turn BPM data-based methods [1, 110, 84]. The ICA-based fitting method was found to have a comparable accuracy with LOCO in the NSLS-II study [110].

In one-pass systems, the betatron phase advance-based methods cannot be applied as there is no temporal betatron oscillation. The method of directly fitting the beam position data to the lattice model can be used in this case [61]. It has been demonstrated with experiments on the LCLS [128].

## CHAPTER 6

Coupling and nonlinear dynamics correction

CONTENTS

The linear coupling between the two transverse planes is caused by the skew quadrupole components in the lattice, which may come from rolls of quadrupole magnets, vertical orbit offsets in sextupoles, random magnetic field errors, or insertion devices. Linear coupling can be measured with BPMs, through either the orbit responses of corrector magnets or the turn-by-turn orbits of a beam undergoing coherent betatron oscillations. Quantities that characterize the linear coupling can be derived from the BPM data and are fitted to the lattice model to determine the linear coupling error sources. The methods for linear coupling measurements include fitting (the off-diagonal blocks of) the orbit response matrix, fitting the amplitudes and phase advances of the normal modes in the cross-plane, fitting the resonance driving terms for the sum and difference resonances, and direct fitting of the turn-by-turn BPM data. The coupling errors can be corrected with skew quadrupole magnets. Skew quadrupole magnets are often coils on multi-function magnets.

In electron storage rings the linear coupling contributes to the equilibrium vertical emittance. Another source of the vertical emittance is the spurious vertical dispersion, which can be generated by vertical bending or coupled from the horizontal dispersion. Sources of vertical bending include the vertical correctors and vertical orbit offsets in quadrupoles. Skew quadrupoles and vertical orbit offsets in sextupoles located in dispersive sections couple the horizontal dispersion to the vertical plane. The vertical dispersion can be corrected with skew quadrupole magnets located in dispersive regions.

Because the vertical dispersion and the linear coupling both contribute to the vertical emittance and are both corrected with skew quadrupoles, the corrections of the two are often combined and are referred to as vertical emittance correction or coupling correction.

Errors in a storage ring lattice also cause the nonlinear beam dynamics behavior to deviate from the design, resulting in a reduction of the dynamic aperture and the local momentum apertures. It would be ideal to correct the errors and restore the nonlinear beam dynamics performance.

In this chapter we discuss the methods of coupling correction. Ideas for nonlinear beam dynamics correction are also discussed.

6.1 COUPLING MEASUREMENT AND CORRECTION METHODS

6.1.1 Off-diagonal blocks of orbit response matrix

If there is no linear coupling, the one-turn 4×4 transfer matrix at a location in a circular accelerator consists of two diagonal 2×2 blocks that represent the motion of the two transverse planes, respectively. The off-diagonal blocks are all zeros. Accordingly, the closed-orbit response of an orbit corrector is limited in the plane of the kick. However, when linear coupling error sources are introduced to the lattice, the one-turn transfer matrix will have non-zero off-diagonal blocks. The values of the off-diagonal block elements are related to the strengths and locations of the error sources. Consequently, the off-diagonal blocks of the orbit response matrix, $\mathbf{R}_{xy}$ and $\mathbf{R}_{yx}$ (Eq. (4.12)), will also be non-zero and the values of these blocks are connected to the error sources.

By fitting the skew quadrupole parameters in the lattice model to minimize the differences between the off-diagonal blocks of the measured and model orbit response matrices, a lattice model that reproduces the linear coupling behavior of the machine can be obtained [102]. The skew quadrupole parameters can be the gradients of actual skew quadrupoles. Corrections can be applied to these magnets to compensate the coupling error sources, resulting in a reduction of linear coupling in the machine. The fitting parameters can also be the rolls of the normal quadrupole magnets. This is equivalent to fitting skew quadrupole components at the locations of the normal quadrupoles. If large rolls on quadrupoles are found, the magnets could be realigned.

The rolls of corrector magnets and BPMs also cause cross-plane orbit responses in the measurements. These effects are not related to the linear coupling in the lattice. Because the orbit response patterns of the corrector and BPM rolls are different from those of the skew quadrupole components, these parameters can be determined by fitting.

In practice the off-diagonal orbit responses are almost always fitted together with the diagonal blocks such that the linear optics and coupling errors are determined simultaneously. This is appropriate as linear optics errors have an impact over the effects of the coupling sources, and vice versa.

By including the measured vertical dispersion data in the fitting objective χ^2, a skew quadrupole setting that produces both the measured vertical dispersion and the cross-plane orbit responses can be determined. Both types of errors can be corrected simultaneously after applying the corresponding corrections to the skew quadrupoles. The weight of the vertical dispersion in the χ^2 definition can be increased in order to achieve the desired level of accuracy of dispersion control.

6.1.2 Amplitude and phase of normal modes via ICA

With linear coupling, the coherent beam motion on the transverse planes can be decomposed into two normal modes. In the case of weak coupling, one of the normal modes can be identified as the horizontal betatron mode and the other the vertical mode. The normal modes will be present on the measured turn-by-turn orbits on both planes. The vertical betatron mode on the horizontal plane and similarly the horizontal mode on the vertical plane at any location are related to the off-diagonal blocks of the one-turn transfer matrix, which in turn are related to the coupling error sources. Using the ICA method to analyze the turn-by-turn BPM data, the normal modes can be separated and their phases and amplitudes at each BPM location can be calculated. By comparing the measured phases and amplitudes to the predictions made by the lattice model, the coupling error sources can be determined and corrected.

At one BPM location, the coupled beam motion on the horizontal and vertical readings can be separated into two pairs of ICA modes, which can be written as [51, 125],

$$x_n = A \cos \Psi_{1n} - B \sin \Psi_{1n} + c \cos \Psi_{2n} - d \sin \Psi_{2n}, \qquad (6.1a)$$

$$y_n = a \cos \Psi_{1n} - b \sin \Psi_{1n} + C \cos \Psi_{2n} - D \sin \Psi_{2n}, \qquad (6.1b)$$

where n is the turn number, $\Psi_{1n} = 2\pi\nu_1 n + \psi_1$, $\Psi_{2n} = 2\pi\nu_2 n + \psi_2$, $\nu_{1,2}$ are the tunes for the two normal modes, mode 1 and 2 are the horizontal and vertical betatron motion, respectively. The phase offsets $\psi_{1,2}$ are common to all BPMs as they share the same temporal patterns.

The coupled motion can be predicted by the 4×4 one-turn transfer matrix at the BPM, $\mathbf{T}$, via $\mathbf{X}(n) = \mathbf{T}^n \mathbf{X}_0$, with the initial phase space coordinate $\mathbf{X}_0$. Matrix $\mathbf{T}$ can be block diagonalized in the form

$$\mathbf{T} = \mathbf{V}\mathbf{U}\mathbf{R}_4(\nu_1, \nu_2)\mathbf{U}^{-1}\mathbf{V}^{-1}, \qquad (6.2)$$

where $\mathbf{V}$ is the same as in Eq. (2.58), and $\mathbf{U}$ and $\mathbf{R}_4$ are

$$\mathbf{U} = \begin{pmatrix} \mathbf{B}_a & \mathbf{0} \\ \mathbf{0} & \mathbf{B}_b \end{pmatrix}, \quad \mathbf{R}_4(\nu_1, \nu_2) = \begin{pmatrix} \mathbf{R}(2\pi\nu_1) & \mathbf{0} \\ \mathbf{0} & \mathbf{R}(2\pi\nu_2) \end{pmatrix}, \qquad (6.3)$$

with $\mathbf{B}$ and $\mathbf{R}$ as defined in Eq. (1.60). The beam motion can then be related to the normal mode coordinates, $\boldsymbol{\Theta}$, through [83]

$$\mathbf{X}(n) = \mathbf{P}\boldsymbol{\Theta}(n), \quad \text{with } \boldsymbol{\Theta}(n) = \begin{pmatrix} \sqrt{2J_1} \cos \Phi_1(n) \\ -\sqrt{2J_1} \sin \Phi_1(n) \\ \sqrt{2J_2} \cos \Phi_2(n) \\ -\sqrt{2J_2} \sin \Phi_2(n) \end{pmatrix}, \qquad (6.4)$$

where $\mathbf{P} \equiv \mathbf{VU}$, $\Phi_{1,2}(n) = 2\pi\nu_{1,2}n + \phi_{1,2}$, and $J_{1,2}$ and $\phi_{1,2}$ are action and angle coordinates for the two normal modes as determined by the initial conditions. The values of $\phi_{1,2}$ at different BPMs differ by the phase advances of the normal modes.

By comparing Eq. (6.1) and (6.4), it can be seen that the amplitudes of the ICA modes and predicted motion are related by [51, 125]

$$\sqrt{A^2 + B^2} = \sqrt{2J_1}p_{11}, \quad \sqrt{c^2 + d^2} = \sqrt{2J_2}\sqrt{p_{13}^2 + p_{14}^2}, \qquad (6.5a)$$

$$\sqrt{C^2 + D^2} = \sqrt{2J_2}p_{33}, \quad \sqrt{a^2 + b^2} = \sqrt{2J_1}\sqrt{p_{31}^2 + p_{32}^2}, \qquad (6.5b)$$

where p_{ij} are elements of the $\mathbf{P}$ matrix and the phase coordinates give

$$\tan^{-1}\frac{B}{A} = \text{Mod}_{2\pi}(\phi_1), \quad \tan^{-1}\frac{d}{c} = \text{Mod}_{2\pi}(\phi_2 + \tan^{-1}\frac{p_{14}}{p_{13}}), \qquad (6.6a)$$

$$\tan^{-1}\frac{b}{a} = \text{Mod}_{2\pi}(\phi_1 + \tan^{-1}\frac{p_{32}}{p_{31}}), \quad \tan^{-1}\frac{D}{C} = \text{Mod}_{2\pi}(\phi_2), \qquad (6.6b)$$

where $\tan^{-1}$ gives function values in the range of $[0, 2\pi)$ using the nominator and denominator of its argument, $\text{Mod}_{2\pi}(\cdot)$ denotes modulo of 2π and we have dropped the initial phase offsets by requiring the phases of the primary modes on the first BPM to be equal between the measured and model values. In Eqs. (6.5)-(6.6), equations relating A, B, C, and D are for the primary normal modes of the two planes, which represent the linear optics. Equations relating a, b, c, and d are for the secondary modes coupled from the other plane, which contain the information of linear coupling.

The differences between the two sides in the equations in Eqs. (6.5)-(6.6) can be characterized by a χ^2 function to be minimized with the least-square method. The action variables, $J_{1,2}$, may be determined by requiring the average beta function values determined from the primary modes for the two planes to be equal to the corresponding model values. Instead of comparing the phase advances differences, their sine and cosine values are compared as it helps eliminate the potential problems due to the discontinuity in the modulo calculation. Horizontal and vertical dispersion functions are also included in the χ^2 function. Each data type can be given a weight factor, in addition to the normalization factor by the corresponding error sigma. The χ^2 function is defined by

$$\chi^2 = \sum_{i,k} \frac{w_k^2}{\sigma_{ik}^2}(d_{ik}^m - d_{ik}^c)^2, \qquad (6.7)$$

where i is the BPM index, k is the data type, and w_k is the weight factor for the data type. There are 14 data types (4 amplitude functions, 8 phase advance related values, 2 dispersion functions) for each BPM, covering the linear optics, linear coupling, and dispersion errors.

The fitting parameters in the lattice model are the quadrupole gradients and skew quadrupole gradients. BPM gains and rolls (or coupling coefficients) are fitted as they are used to scale and rotate the spatial vector elements with Eq. (4.17) or Eq. (4.18). For example, the (A, a) pair will be modified with

$$\begin{pmatrix} \tilde{A} \\ \tilde{a} \end{pmatrix} = \begin{pmatrix} g_x & c_x \\ c_y & g_y \end{pmatrix}^{-1} \begin{pmatrix} A \\ a \end{pmatrix}. \tag{6.8}$$

The same transformation applies to (B, b), (c, C), and (d, D) pairs.

As the linear optics and coupling least-square fitting problem typically suffers from the degeneracy difficulty due to the similarities between the effects of the fitting parameters, the constrained fitting method is needed to find a reasonable solution that can be used for correction [50].

6.1.3 Other methods of coupling correction

The linear coupling information can also be uncovered from the turn-by-turn BPM data with other data analysis methods.

Direct fitting of turn-by-turn BPM data: The direct fitting of turn-by-turn BPM data discussed in Section 5.3.2 can simultaneously determine the linear optics and coupling errors. The fitting setup only needs to be modified to add BPM rolls (or coupling coefficients) and skew quadrupole gradients as fitting parameters. This approach has been demonstrated in simulation with the SPEAR3 lattice model [61].

Resonance driving terms: The cross-plane motion can be characterized by the resonance driving terms for the linear sum and difference resonances. Correction of the resonance driving terms should lead to the correction of the linear coupling [39, 47].

The RDTs that drive the linear difference and sum resonances are f_{1001} and f_{1010}, respectively, which are related to skew quadrupole errors through

$$f_{\mp} = \frac{\sum_k \Delta a_{1,k} L_k \sqrt{\beta_{x,k} \beta_{y,k}} e^{i(\psi_{x,k} \mp \psi_{y,k})}}{4(1 - e^{i2\pi(\nu_x \mp \nu_y)})}, \tag{6.9}$$

where $f_- \equiv f_{1001}$ drives the linear difference resonance, $f_+ \equiv f_{1010}$ drives the linear sum resonance, Δa_1 and L are the gradients and lengths of the skew quadrupole error sources, respectively. The corresponding spectral lines for f_{1001} are ν_y on h_x^- and ν_x on h_y^-, respectively. The spectral lines for f_{1010} are $1 - \nu_y$ on h_x^- and $1 - \nu_x$ on h_y^-, respectively.

Using a pair of adjacent BPMs to derive the angle coordinates, the turn-by-turn resonance basis coordinates, $h_{x,y}^-$, can be calculated, from which the linear coupling RDTs can be determined. Knowing the amplitudes and phases of the linear coupling RDTs, the skew quadrupole errors in the lattice model can be fitted with the least-square method, using Eq. (6.9) to calculate the model RDT values. The measured vertical dispersion function at the BPMs also needs to be included in the fitting input data in order to reduce the vertical emittance contribution from the vertical dispersion.

Transfer matrices: With the full phase space coordinates determined from two adjacent BPMs, the one-turn transfer matrices at the BPMs can be calculated with Eq. (4.8) or by fitting parameters to construct a symplectic transfer matrix. The two BPMs can be separated by a simple drift space, or a short section with a few magnets. In the latter case, the model transfer matrix between the two BPMs can be used to calculate the angle coordinates, x' and y' (Eq. (4.5)).

The one-turn transfer matrices contain both linear optics and coupling information. The lattice model can be fitted by minimizing the differences between the measured and model one-turn transfer matrices at multiple BPM locations. The skew quadrupole gradients are the fitting parameters for the lattice model. BPM calibration errors and rolls are also included as fitting parameters.

6.2   COUPLING CORRECTION EXPERIMENTS

Coupling correction with orbit response matrix has become a standard practice at many electron storage rings. Typically the vertical emittance can be corrected to the level of a few picometer (pm)-rads. In some cases, the vertical emittance has reached the sub-pm level [30].

In the NSLS-II optics correction experiment discussed in the previous chapter, BPM coupling coefficients and skew quadrupoles were also fitted for the data sets before and after the corrections (using the ICA method) were applied to the machine. Figure 6.1 shows the ratio of the rms orbit response in the vertical plane to that of the horizontal plane due to horizontal correctors (H), or similarly the ratio of horizontal responses to vertical responses due to vertical correctors (V), before and after the corrections were applied. The average value of such ratios was reduced from 0.054 to 0.025 by the corrections.

Linear coupling is also observed in the turn-by-turn BPM data. A direct measure of the level of coupling is the amplitude ratios of the normal modes in the cross-plane and in the primary plane, defined as

$$r_1 = \frac{\sqrt{a^2 + b^2}}{\sqrt{A^2 + B^2}}, \qquad r_2 = \frac{\sqrt{c^2 + d^2}}{\sqrt{C^2 + D^2}}. \tag{6.10}$$

Figure 6.2 shows the r_1 and r_2 ratios for the turn-by-turn BPM data taken before and after corrections in the same experiment as in Figure 6.1. The average ratios were reduced from $r_1 = 0.094$ and $r_2 = 0.107$ to 0.046 and 0.052, respectively.

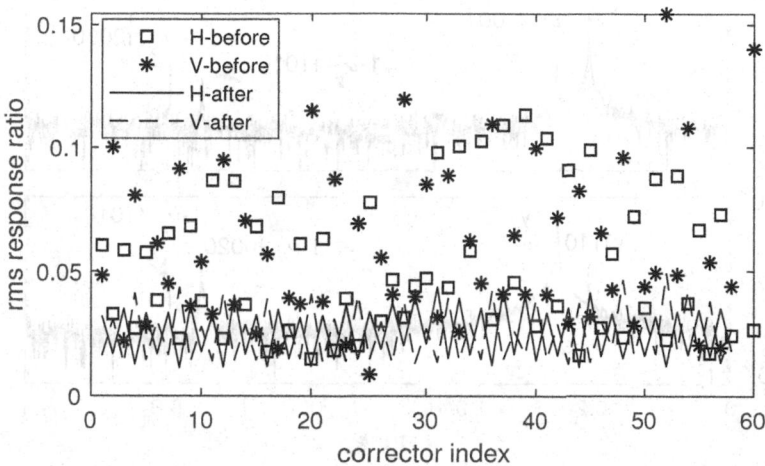

Figure 6.1 The ratios of cross-plane and in-plane rms orbit responses for the correctors, before and after the ICA-based optics and coupling corrections were applied in an NSLS-II experiment.

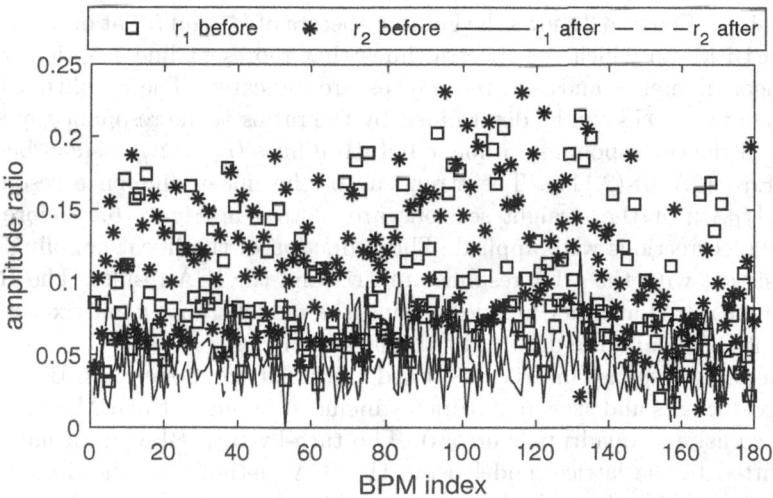

Figure 6.2 The amplitude ratios of betatron normal modes (obtained by ICA) in the cross-plane and the primary plane (see Eq. (6.10)) before and after corrections were applied in the NSLS-II experiment.

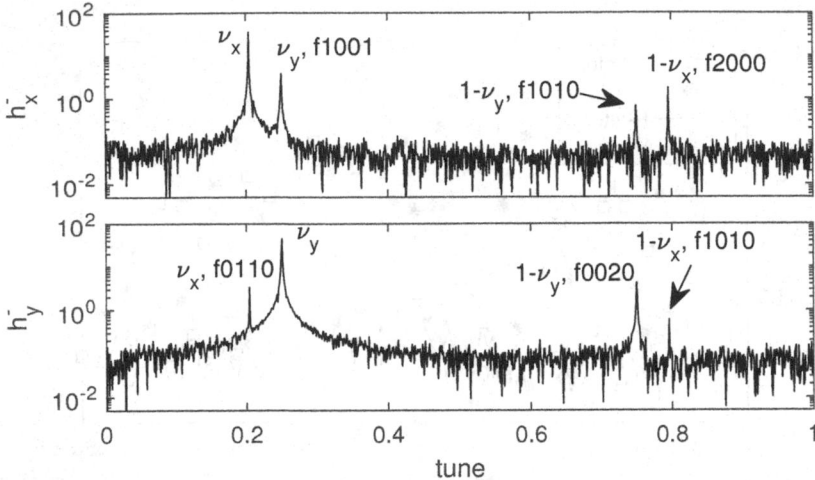

Figure 6.3 An example of resonance driving term identification on the Fourier spectra of turn-by-turn resonance basis coordinates. Top: horizontal; bottom: vertical.

There are two BPMs located in each of the straight sections in NSLS-II. The full phase space coordinates can be calculated from the turn-by-turn BPM data at these locations, from which the resonance basis coordinates can be obtained. Figure 6.3 shows the Fourier spectra of h_x^- and h_y^- at one straight-section BPM, on which the spectral lines that represent linear optics errors, the linear difference and sum resonances are indicated. The amplitudes and phases of the RDTs can be determined by the ratios of the resonance spectral lines and the corresponding primary betatron lines (ν_x or ν_y), as can be seen from Eqs. (2.116)-(2.117). The strengths of the linear difference resonance RDT, $|f_{1001}|$, at the straight sections are plotted in Figure 6.4 before and after the corrections were applied. The reduction of the linear coupling level is consistent with the orbit response matrix and the ICA results. The RDTs can be measured at other locations using the model transfer matrix between adjacent BPMs and are used to fit the lattice model.

The orbit response matrix was fitted to the lattice model with BPM coupling coefficients and skew quadrupoles included (along with BPM gains, corrector gains, and quadrupole errors). The turn-by-turn BPM data have also been fitted to the lattice model using the ICA method and the direct orbit fitting method. Both methods include BPM gains and quadrupole errors in the fitting parameters. The fitted integrated skew quadrupole gradients for the three methods are shown in Figure 6.5. While there are some differences in the fitted skew gradients, the linear coupling ratios of the resulting fitted lattices are very similar, which are 1.29% (LOCO), 1.50% (ICA), and 1.42% (fit X/Y) for the three methods, respectively. The distribution of the vertical

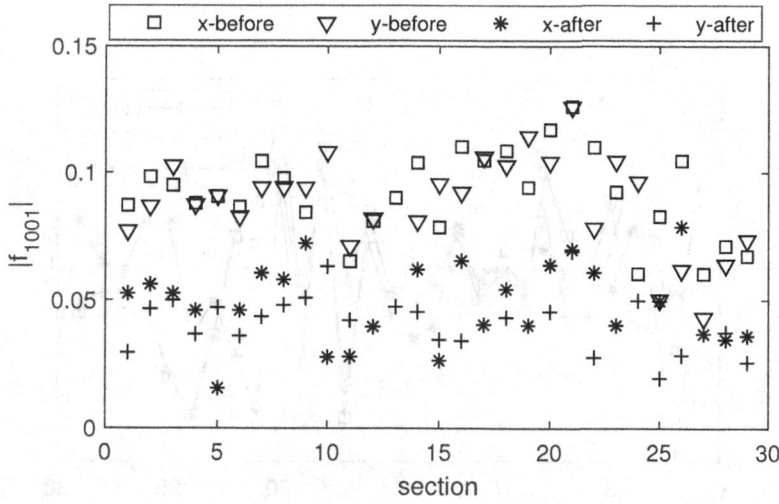

Figure 6.4 Strengths of the linear difference resonance spectral line, $|f_{1001}|$, from turn-by-turn BPM data on both planes (x or y) before and after corrections.

emittance around the ring for the three fitted lattice are shown in Figure 6.6. The corrections with the ICA result led to a reduction of the coupling level as seen in the model independent measures shown in Figures 6.1, 6.2, and 6.4, as well as in the fitted lattices. The coupling ratio was corrected to about 0.3%.

6.3 CORRECTION OF NONLINEAR DYNAMICS ERRORS

Correction of linear errors in accelerators, i.e., orbit errors, optics errors, and linear coupling, has generally been successful on accelerators equipped with modern diagnostics. This may be adequate for many applications, such as linacs, transport lines, and some synchrotrons.

However, as storage rings keep pushing toward lower emittances, it has become a big challenge to ensure the rings have sufficient nonlinear beam dynamics performance. A large dynamic aperture is required for the injection of beams into the ring and large local momentum acceptances are needed for a long beam lifetime for high current beams. Large rings with low emittances tend to have small horizontal dispersion and large natural chromaticities (negative). Chromaticity correction requires more and stronger sextupole magnets, and consequently, the lattices become more nonlinear and their dynamic apertures become smaller. The design of the sextupole scheme is a critical component of a low emittance rings lattice. Typically, the design lattice achieves an acceptable dynamic aperture and momentum apertures only after an extensive numeric optimization, using multi-objective genetic algorithms (MOGA) [16, 122] or particle swarm optimization (PSO) [59].

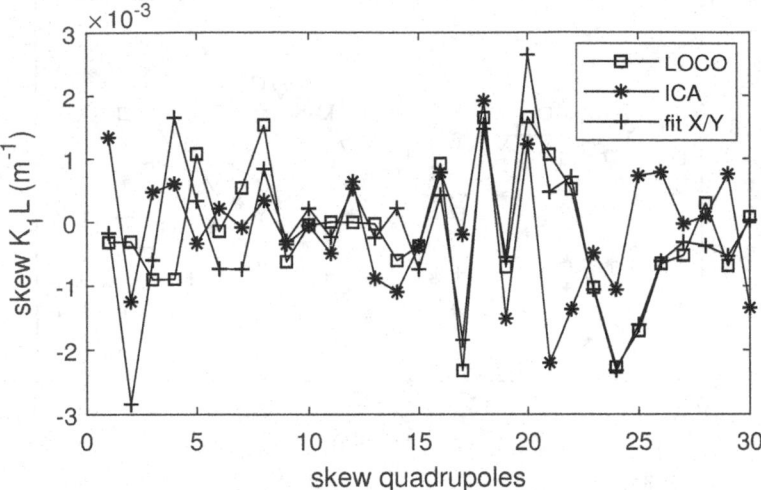

Figure 6.5 Integrated skew quadrupole gradients in the NSLS-II experiment before corrections by fitting the orbit response matrix (LOCO), fitting the ICA results (ICA), or directly fitting turn-by-turn BPM data (fix X/Y).

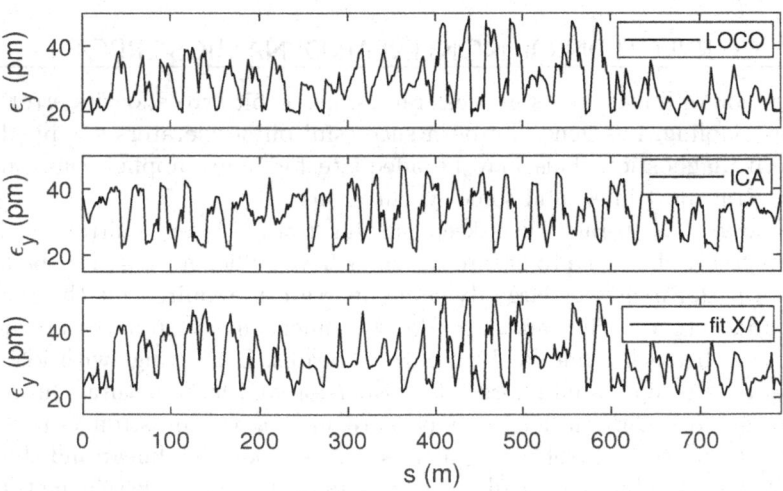

Figure 6.6 The vertical projected emittances throughout the ring in the lattices fitted by the three methods.

Realizing the design performance of the nonlinear beam dynamics on the real machine is also very challenging and will be more challenging for the future diffraction limited storage rings as these rings will be more nonlinear and have more error sources. Developing methods to compensate the negative impact of lattice errors to the nonlinear beam dynamics is critical for the success of future storage rings and can also benefit the operation of existing rings.

The nonlinear beam dynamics performance of a storage ring depends strongly on the linear optics. Correction of linear errors is essential. For third generation light sources, optics correction often leads to improvements in the dynamic aperture and the momentum acceptances. However, there will always be residual optics errors. Even when the beta functions and phase advances at the BPMs are at the ideal design values, their values at the sextupoles may differ from the design. The strengths of the nonlinear magnets in the real machine may be different from the design model due to calibration errors. The nonlinear fields in dipole and quadrupole magnets and the higher order multipoles in sextupoles due to systematic and random errors may not be included in the model. Therefore, correction of the nonlinear lattice features toward the design is necessary.

Nonlinear beam dynamics measurement and correction:
Naturally, it is desirable to apply the beam-based correction approach to correct the nonlinear beam dynamics toward the design. As for the cases of linear errors correction, this requires

- control parameters (i.e., knobs) that have a direct impact to the nonlinear dynamics,

- beam diagnostics that can effectively measure the nonlinear beam dynamics behavior, and

- a method that deduces the required adjustments of the knobs from the measurements.

The knobs are typically the strengths of the nonlinear magnets in the lattice, i.e., sextupoles and octupoles (if present in the lattice). The fitting results for these knobs can be used to correct the machine. If the goal is to identify sources of discrepancies between the model and the machine, additional multipole components in the model (such as sextupole, octupole, and decapole components of the dipole and quadrupole magnets) can be included as fitting parameters. The diagnostics need to detect features of the beam dynamics that are relevant to the nonlinear beam dynamics performance and can be compared to the design model. These features could be higher order chromaticities, the tune shifts with amplitudes, and the resonance driving terms (RDTs). The least-square fitting approach can be used to calibrate the lattice model with the measurements [9, 8].

Tune shifts with amplitudes and RDTs can be measured with turn-by-turn BPMs. As shown in Eqs. (2.116)-(2.117), the spectral lines of the turn-by-turn orbits are related to the RDTs. Each RDT drives a specific resonance and corresponds to a specific spectral line. Conversely, each resonance is driven by many RDTs, which may come from different multipole fields through various perturbation mechanisms. The strengths of the nonlinear RDTs depend on the oscillation amplitudes to various orders of power according to the orders of the RDTs. To effectively sample the nonlinear fields, the beam needs to be excited to coherent oscillations with a relatively large amplitude. However, when measuring the lower order RDTs, it is desirable to suppress the contributions of the higher order terms by using a moderate amplitude. In theory, it is possible to determine the lower order RDTs and in turn the corresponding generating function terms, which are then used to remove the lower order contributions to the spectral line in subsequent experiments with larger amplitudes, allowing the separation of the RDTs order by order [6]. In practice this is very difficult as there would be considerable errors even in the lowest order RDTs.

In experiments, it has been demonstrated that the leading order RDTs by sextupoles and octupoles can be determined from the measured turn-by-turn BPMs [40]. The spectral lines on the horizontal and vertical turn-by-turn resonance basis orbits (i.e., h_x^- and h_y^-) corresponding to the leading order RDTs for normal sextupoles are listed in Table 6.1. For example, the f_{3000} coefficient drives a spectral line (-2, 0) on the horizontal orbit, whose fractional tune is $1 - 2\nu_x$ and the amplitude is $12I_x|f_{3000}|$. The phase of the RDT coefficient $\phi_{jklm} = \text{Arg} f_{jklm}$ can also be determined as the phase of the spectral line is simply $\phi_{jklm} + \psi_{x/y,0} - \pi/2$, where $\psi_{x/y,0}$ is the phase of the betatron tune line on the corresponding plane [10].

The strengths of the leading order RDTs by sextupoles are linearly proportional to the integrated gradients of the sextupoles. For example, the f_{3000} coefficient is related to sextupole strengths via

$$f_{3000} = -\frac{\sum_i b_{2,i} L_i \beta_{x,i}^{\frac{3}{2}} e^{i3\Delta\phi_{x,i}}}{48(1 - e^{i6\pi\nu_x})},$$ (6.11)

where $b_{2,i} L_i$ is the integrated strengths of sextupole i and $\Delta\phi_{x,i}$ is the hori-

TABLE 6.1 Identification of leading order RDTs driven by normal sextupoles on spectral lines of horizontal and vertical turn-by-turn orbit. The HSL and VSL show the locations of the spectral lines and $|H/f_{jklm}|$ and $|V/f_{jklm}|$ are the ratios of the corresponding amplitudes over the f_{jklm} coefficients.

f_{jklm}	HSL	$|H/f_{jklm}|$	VSL	$|V/f_{jklm}|$
f_{3000}	(-2,0)	$12I_x$	N/A	
f_{1200}	(2,0)	$4I_x$	N/A	
f_{1020}	(0,-2)	$4I_y$	(-1, -1)	$8\sqrt{I_x I_y}$
f_{0120}	N/A		(1, -1)	$8\sqrt{I_x I_y}$
f_{0111}	N/A		(1, 1)	$4\sqrt{I_x I_y}$

zontal phase advance between the sextupole and the observation point. The phase of the complex RDT coefficient changes with location in the ring but its amplitude does not change except at the locations of the multipole magnets that drive the resonance. Across a multipole magnet, the RDT coefficient has a step change as the contribution from the multipole magnet changes phase. If the RDTs can be measured before and after the multipole magnet, the strength of the magnet could be determined directly.

In reality, there are not so many observation points to directly measure the strengths of all multipole magnets. However, the connections between the RDTs and the multipole magnets provide a way to resolve the multipole errors. This can be done by fitting the measured RDTs to the lattice model. Other observed features of the nonlinear dynamics can also be included. In general, the objective function may be defined as

$$
\chi^2 = f(\mathbf{p}) = \sum_i^{2M} w_k^2 \sum_{k \in \Phi} (f_{i,k}^{(m)} - f_{i,k}^{(c)(\mathbf{p})})^2, \tag{6.12}
$$

where $\mathbf{p}$ contains all fitting parameters, the summation of i is over all horizontal and vertical BPMs, k represents a feature parameter in Φ, the collection of selected features (e.g., RDTs), w_k is the weight assigned for the feature, and superscripts (m) and (c) stand for measurements and calculations, respectively. Global parameters, such as detuning coefficients, and chromaticities, and higher order chromaticities, can also be included in Eq. (6.12).

The determination of the resonance basis coordinates needs to use a pair of adjacent BPMs and the lattice errors between the two BPMs can introduce errors to the angle coordinates. To avoid this complication, the spectral lines of the turn-by-turn positions can be used directly to fit the nonlinear lattice model. In this case the spectral lines are typically combined effects of two or more RDTs. Details of the combined RDTs for the typical spectral lines can be found in Ref. [40].

The nonlinear response of BPMs to beam positions due to the geometric configuration of the buttons as discussed in Chapter 3 has a significant impact to the nonlinear dynamics measurements. Signal processing in the BPM electronics that produces the turn-by-turn position may involve samples from multiple turns. This will change beam position reading from the actual value and need to corrected before the data are used for the physics analysis [8].

Challenges for nonlinear lattice correction:
The method of nonlinear lattice correction with RDTs faces many serious practical challenges. First, beam decoherence due to chromaticity and nonlinear detuning may limit the number of turns of usable data and in turn the precision of RDT measurements. While chromaticity can be easily set to zero, it is usually not easy to simultaneously set nonlinear detuning to zero. Even if it can be done, the nonlinear lattice would have been changed so much from the design lattice such that the correction may not be useful for the operation of the machine. Nonlinear detuning will become substantially more severe in

the diffraction limited storage rings, making it more difficult for the method to work where it is needed the most. Radiation damping in high energy storage rings also limits the number of usable turns.

Second, the imperfections of the BPMs will affect the RDT measurements. These include the nonlinear response of the BPMs to beam position and the random BPM noise or other contaminating noise sources. The nonlinear BPM response distorts the apparent nonlinear beam motion and interferes with the spectral lines. The random BPM noise introduces errors to the RDTs or even completely swamps the weak spectral lines. The decoherence and BPM imperfections affect the precision of RDT measurements. The errors in the first order RDTs driven by sextupoles could be high enough to make the results not useful for correction. The relative errors in higher order RDTs originated from interactions of multiple sources would be much larger.

Third, there are not enough independent knobs for the correction of the many RDTs. In theory, if linear and nonlinear lattice elements in the machine are identical to the design model, RDTs of all orders are automatically equal between the machine and the model. In reality, there are residual optics errors in the machine and there are also physics effects not accounted for in the model (such as fringe fields, insertion device perturbations, etc.) and hence the RDTs will be different. Compensation of all the relevant RDTs will likely require more nonlinear magnets than are available.

Another issue for the RDT-based correction approach is that the correction of measurable RDTs would not necessarily bring improvement to the nonlinear dynamics performance. The dynamic aperture of a storage ring may be limited by one or more higher order resonances which are not visible on the BPM signals. For example, one limiting resonance in the APS-U lattice is the 6th order resonance $2\nu_x + 4\nu_y = p$. The correction may likely introduce changes to the lattice such that other resonances become performance limiting. Furthermore, chromatic resonances are difficult to measure and correct and hence the local momentum apertures may not be under control.

III

Beam-based Optimization

Online optimization algorithms

CONTENTS

Beam-based optimization is the approach of adjusting the control parameters of the accelerator to optimize its performance, using the real-time, measured beam performance as the guide for choosing the trial settings. Manual tuning is a type of beam-based optimization in its original form. While manual tuning is an indispensable approach and is widely used, its application is usually limited to simple, small-scale problems. Automated tuning is an advanced form of beam-based optimization, in which the computer takes the role of a human being to make decisions. By closely interacting with the control system to dial in the trial solutions and to perform analysis on the diagnostic data and employing advanced online optimization algorithms, automated tuning can deal with bigger and more complex problems. It not only can expedite the routine tuning processes, but more importantly, can solve accelerator tuning problems that cannot be accomplished with other methods.

Online optimization algorithms are essential to the strength of automated tuning. Because of the measurement errors that enter the performance metrics and other characteristics of the online application environment, online optimization algorithms need to address special challenges. Traditional optimization algorithms that are powerful for smooth functions might not be suitable for online problems. New algorithms that were modified from traditional algorithms with simple-minded improvements may be powerful tools.

In this chapter we will first discuss the general considerations of the online optimization approach. This is followed by a review of the optimization algorithms for continuous functions. Analytic functions with added random errors are used to test the performance of the algorithms for online applications. The robust conjugate direction search (RCDS) method is introduced and compared to the other algorithms through the tests.

7.1 GENERAL CONSIDERATIONS OF ONLINE OPTIMIZATION

7.1.1 Need for online optimization of accelerators

Beam-based correction is a satisfying approach. It seems natural to accelerator physicists as it employs the physics principles that govern the machines. The correction process is deterministic. The success of the correction approach signifies that accelerator physicists understand and have control over the inner workings of the machine.

However, the correction approach is not always applicable. For some machines or some applications, there are no diagnostics available to detect the discrepancies between the machine and the ideal target. In some cases, the diagnostics cannot provide the detailed information needed to solve for the required parameter adjustments. In other cases, the diagnostics simply do not exist. There may be a lack of a target for correction. If there are major differences between the model and the real machine, the target suggested by the model may not be applicable. In this case, the target itself needs to be discovered. It is also possible that both diagnostics and the target exist, but there is no reliable, deterministic method to deduce the solution for the correction toward the target. In these circumstances, empirically tuning the control parameters for the optimal performance is a sensible choice.

Manual tuning is frequently employed in accelerator control rooms by accelerator physicists and operators. During manual tuning, one or a few control parameters (i.e., knobs) are adjusted while the machine performance is monitored. Depending on the applications, the machine performance indicator may be beam intensity, transmission or capture efficiency, beam sizes, beam loss, beam lifetime, etc. Guided by the response of the machine performance to the knob changes, the human tuner makes further knob adjustments in order to maximize the machine performance.

Manual tuning is essentially an optimization process. The function to be optimized is the machine performance evaluated on the operating machine through measurements. The knobs are the input variables of the function. The human tuner executes an optimization algorithm to search the parameter space for the optimum of the performance function. Manual tuning has many limitations. It is typically slow for humans to dial in the new setpoints, to process the measured data, and to make decisions on the next move. The complexity of the optimization problem is usually limited by the ability of humans to analyze and comprehend the data taken from a high dimension parameter space in real time. Typically only one knob is tuned at a time. Efficient search directions that involve multiple knobs cannot be taken advantage of. It is not practical to tune problems with a large number of knobs. Small trends over a long parameter range could be overlooked, which may prevent the convergence toward the optimum in some cases. The successful execution of the manual tuning depends on the tuner's experience, training, and familiarity with the system. This could pose a challenge to the training and retaining of a competent operation team.

In the age of computerized accelerator control, it is clearly desirable to automate the tuning process. Automated tuning integrates all the three components – knob adjustments, performance monitoring, and decision making – in one computer program. This not only speeds up the data taking and the data processing, but also opens up the possibility of using efficient optimization algorithms. Optimization of large scale problems with complex parameter space (e.g., with strong coupling between the parameters) becomes feasible. Simultaneous optimization of multiple performance measures is also possible.

Automated tuning has been attempted at many places [35, 32, 34, 2]. The algorithms employed in these studies include one-dimensional scans, random optimization, Nelder-Mead simplex, etc. Despite these efforts, automated tuning has not gained much popularity until recently. This was likely due to the challenges of online optimization that were not met by the traditional optimization algorithms. The power of automated tuning will be manifested only when algorithms suitable for online optimization are adopted. There is an ongoing campaign to develop new algorithms for the online tuning of accelerators. As more research is carried out in the area, more effective and efficient algorithms will appear, which may change the landscape in this emerging field.

7.1.2 Formulating online optimization problem

Online optimization is similar to ordinary mathematical optimization in that it looks for the maximum or minimum of the objective function(s) within a certain parameter space. We consider the minimization problem only as any maximization problem can be turned into minimization by flipping the sign

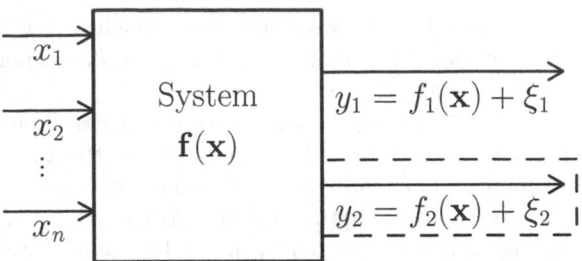

Figure 7.1 In beam-based optimization, the system is a black box that evaluates the objective function(s) using the input variables.

of the objective function. Unlike the ordinary cases, the objective function is not given in an analytic form or calculated through a computer program. Instead, the function is evaluated through measurements on a machine. The relevant operating conditions of the machine are controlled through the input variables; all other conditions either have no impact to the objective function or remain unchanged during the course of the optimization. In other words, a set of input variables uniquely determine the values the objective functions, apart from the inevitable random measurement errors. Knowing the working principle of the machine is not essential to online optimization. The system to be optimized can be considered as a black-box, as illustrated in Figure 7.1.

Normalization of parameter range: For a real machine, every tuning parameter has a finite valid range. The numeric ranges for the different input variables can vastly differ, as the parameters are often different physical quantities. Large differences in the scales of parameter ranges can cause numeric difficulties in some cases. Implementation of the optimization algorithms may become complicated by the need to accommodate the scale differences, for example, in terms of assigning the initial step sizes. Therefore, it is sensible to normalize all input variables to a standard range, which we choose to be $[0, 1]$. For a parameter p with the physical range of $[p_{\min}, p_{\max}]$, the conversion between the normalized value, x, and the physical value is simply

$$x = \frac{p - p_{\min}}{p_{\max} - p_{\min}}, \qquad p = p_{\min} + (p_{\max} - p_{\min})x. \qquad (7.1)$$

For a multi-variable problem with n input variables, the normalized parameter space is the n-dimensional unit hypercube. For some applications, the parameter space may be limited through other forms of constraints. For example, a corner of the hypercube may be excluded, or the point representing the vector of input variables, $\mathbf{x} = (x_1, x_2, \cdots, x_n)^T$, may be required to be within a sphere. These cases are not very common in accelerator applications. Such constraints can be enforced through the definition of the objective functions, e.g., by returning unusually large values when the constraints are violated.

Single-objective optimization: Under most circumstances, only one objective function is optimized at a time. This is the case as long as tuning the related control parameters does not negatively impact the machine performance in other measures. With the normalized parameters, the optimization problem for one objective function can be expressed as

$$\mathbf{x}_{min} = \arg\min_{\mathbf{x}\in[0,1]^n} f(\mathbf{x}). \tag{7.2}$$

In accelerator tuning problems, the objective function, $f(\mathbf{x})$, can be considered a smooth function. However, the function value evaluated through measurement, $\bar{f}(\mathbf{x})$, is not smooth, as there is always a random deviation in the measurement from the true performance function,

$$\bar{f}(\mathbf{x}) = f(\mathbf{x}) + \xi, \tag{7.3}$$

where ξ is a random variable that denotes the measurement error.

The random error has a significant impact over the behaviors of the optimization algorithms. Many traditional algorithms assume the objective function to be smooth. Depending on the working principle of the algorithms and the nature of the optimization problems, with noise in the objective function, the algorithms may have slow convergence or fail to converge to the minimum. Robustness against noise is a main requirement for online optimization algorithms.

Multi-objective optimization: There are cases when the machine performance in terms of multiple measures needs to be simultaneously optimized, while the optimal setting for one measure may not be the best condition for the other(s). As a gain made in one measure by adjusting the knobs could lead to the loss in another measure, a trade-off between the performance measures is needed in selecting the operation condition.

A common approach is to combine these measures into one objective by assigning weights to the individual objectives,

$$f_w = \sum_i^M w_i f_i, \tag{7.4}$$

where M is the number of objectives and the weights, w_i, $i = 1$–M, account for both the scale differences of the objectives and their relative importance. The combined objective can also be defined with normalized values

$$f_w = \sum_i^M w_i \frac{f_i - f_i^{target}}{L_i}, \tag{7.5}$$

where f_i^{target} is the target performance for the i'th performance metric, and L_i's are scale constants that bring the metrics to comparable numeric values.

In the above approach the assigned weights in the definition of the combined objective function have a big impact to the optimal solution. This is not ideal since the choice on the weights is made before all possibilities are presented. Sometimes it is desirable to find a distribution of optimal solutions before choosing the solution to operate with. Multi-objective optimization is needed in such cases. The goal here is to find solutions $\mathbf{x}$ in the unit hypercube that simultaneously minimize multiple objective functions,

$$\min(f_1(\mathbf{x}), f_2(\mathbf{x}), \cdots, f_M(\mathbf{x})). \tag{7.6}$$

Comparison of two solutions for the single objective case is straightforward. In the minimization problem that we are considering, the solution with a lower value for the objective is the better one. However, comparison of two solutions in a multi-objective optimization is more complicated since there can be additional outcomes – solution A can be better than solution B in one objective but worse in another. Non-dominated sorting is used to classify the solutions by their performances in this case. Solution A is said to dominate solution B if A is at least equal to B in all objectives and is strictly better than B for at least one objective, i.e.,

$$\begin{aligned} \forall i \in [1, M]: \quad & f_i(\mathbf{x}_A) \leq f_i(\mathbf{x}_B), \text{ and} \\ \exists j \in [1, M]: \quad & f_j(\mathbf{x}_A) < f_j(\mathbf{x}_B), \end{aligned} \tag{7.7}$$

where $[1, M]$ stands for all integers from 1 to M. Using non-dominated sorting, a group of solutions can be ordered in different fronts, from best to worst, labeled, F_1, F_2, $\cdots$, such that any solution in F_i dominates any solution in F_j if $i < j$, but no solution dominates another solution if they are in the same front. The leading front for all valid solutions in the parameter space is called the Pareto front. Solutions in the Pareto front represents the best possible solutions. The goal of multi-objective optimization is to find the Pareto front. The $M = 2$ case is the most common for multi-objective optimization.

Unless noted, in the following we consider single-objective optimization.

7.1.3 Practical considerations for online optimization implementation

Automated online optimization is executed by computer programs. The computer programs need to implement the optimization algorithms, define the optimization problem, provide an appropriate interface with the users and the control systems of the machine, and manage and process the data collected during the optimization process. While the algorithms are common to all applications, the problem setup can be very different from application to application. By utilizing a proper interface between the algorithm implementation and the problem setup in the online optimization program, the users can be shielded from the details of the algorithms and can thus be focused on the particular application. In the same time, the developers of the optimization algorithms can be isolated from the details of the specific applications. Such an interface not only makes the program easy to use for users, but also makes the program easy to maintain, support, and extend.

In the following we describe a general framework that has been demonstrated to be easy to use through many real-life applications. Pseudo code scripts are used as illustrations. Typically a script is used as the main control, in which the initialization of the environment variables, the initial setup of the machine conditions, the launching of the optimization algorithm, and the post-processing may be conducted. The objective function is defined in a separate function. The same principles can be applied for the implementation in other programming environments.

Setup of the optimization problem: In the optimization problem setup, several global variables are defined. These include the number of objective functions, the number of knobs, and the ranges of the knobs. For example, for a single objective function problem with 4 knobs, the variables are defined with

```
Global_Variable Nobj Nvar VRange
Nobj = 1;
Nvar = 4;
VRange = [1, 1, 1, 1]'*[-2,2]; #a 4-by-2 matrix
```

where 'VRange' is a $n \times 2$ matrix, the two numbers on each row of which give the low and high limits of the corresponding knob. In this example the parameter range is set to [-2.0, 2.0] for all knobs. Other parameters that are used in the optimization algorithms or for changing the machine conditions can also be defined here. For example, the noise sigma for the objective function is needed for some algorithms.

The parameter range may be given in terms of the actual limits, or it can be given relative to the initial value of the parameter. In the latter case, the initial value needs to be passed into the objective function, in which it is used to convert the normalized parameter to the physical value. The initial values can be passed as global variables.

The optimization problem is defined through the objective function. The interface of the objective function is

```
Function y = func_obj(x)
```

where 'x' is a vector of the normalized parameter values and 'y' is the return value of the objective function. The function handle will be passed to the optimization algorithm. Inside the objective function, the normalized parameters in 'x' are first converted to the physics parameters with the global variable 'VRange'. The physics parameters are then set to the machine. Typically a pause is needed for the machine to settle to the new condition. For example, it may take a few seconds for a magnet to settle to a new setpoint. The code may also check the readbacks of the parameters and wait until the readbacks are equal to the setpoints within a certain tolerance. After that, the machine performance is measured. This could be as easy as reading a process variable (PV) served by the control system, or the code may need to take data, analyze the data, and derive the objective function value. Multiple readings may be taken for averaging to reduce the noise level, especially if the noise is a big issue and the measurement of the performance is fast relative to the time needed to change the machine conditions. The performance measure is then returned as the value of the objective function.

Inside the objective function, the code may need to monitor the machine conditions for any anomaly that could arise. It should also be able to pause the experiment and alert the user if an action needs to be taken by the user before the experiment can continue. For example, in an injection efficiency optimization experiment, the code needs to stop when the beam has reached the maximum current, at which point the beam needs to be dumped.

Interface to optimization algorithms: The interface between an optimization algorithm and the specific application is the function call to the optimizer. The required information is passed as arguments to the optimizer function, which may be defined as

```
Function [xm,ym,dout]=optimizer(func,x0,MaxEval,OtherInput)
#Input parameters:
#  func, function handle to the objective function
#  x0,   the initial normalized parameter vector
#  MaxEval, maximum number of evaluations
#Output parameters:
#  xm, parameter vector for the optimal solution
#  ym, the corresponding objective function value
```

where 'OtherInput' represents additional input arguments and 'dout' is a structure that holds additional outputs that are specific to the optimizer.

The optimizer searches the n-dimensional unit cube for a solution that gives the minimum value of the objective function, starting from the initial solution 'x0'. The algorithm will keep track of the number of evaluations of the objective function and check it against the 'MaxEval' value frequently (e.g., at the end of each iteration). It will exit the optimizer if the number of evaluations exceeds 'MaxEval'. Other termination conditions can also be implemented.

Depending on the nature of the optimizer, additional input parameters may be needed. Some examples include the noise sigma of the objective function, the initial step size, the initial conjugate direction set, etc.

Data management: During the course of an online optimization run, many data points will be evaluated and at each data point, many machine condition and performance parameters will be recorded. It is desirable to save these data for post processing. Since online optimization often exits prematurely, before a pre-determined termination condition is met, it is essential to store the data frequently so that no loss of data occurs.

A good place to manage the optimization data is in the objective function as this is where data are taken. At each function evaluation, all relevant data can be put in a structure and appended to a data file. A simple way of data keeping is to save data to a global variable. The global variable may be a list of the data structures, one entry for each function evaluation. Or the global variable can be a data array, each row of which is a data entry.

The data variable is reset in the setup script, by, e.g.

```
Global_Variable g_cnt g_data
g_cnt=0; #reset the counter
g_data=[];
```

Inside the objective function, after the measurements are done, the data entry is entered, with

```
g_cnt=g_cnt+1;
g_data=[g_cnt, p(:)', ym, OtherParas];
```

where 'p(:)' is a row vector with all physical values for the knobs and 'OtherParas' represents any other parameters that need to be saved. Time stamp can also be saved for each entry.

The global variable will persist in the workspace even after a forced quit (e.g., with Ctrl+C). After the optimizer is stopped, the data variable can then be saved to a file. The saved data can be processed to find the best solution among all evaluated solutions.

In the above scheme global variables are used to share data between the setup script and the objective function. This should pose no problem because usually the optimization program is small in scale. In the scripting environment it has an advantage in that the data are immediately available for post-processing in the same workspace.

An alternative scheme is to use the object-oriented programming practice, with which all setup and data variables are defined as member variables of a class object, along with the optimizer and data processing functions. The setup variables are specified at the time the object is created or initiated. The objective function is defined as a standalone function, whose handle is passed to the class object at initialization and to be saved as a member variable. A member function of the class serves the role of the objective function for all evaluations internal to the class. Inside this member function the parameter range conversion is done and then the external objective function is called. The parameter ranges need not to be passed outside the object.

Optimization progress monitoring: Monitoring the progress of the optimization process is important for online applications. With real time monitoring, any unexpected or undesired behaviors of the optimization program can be discovered for the program to be terminated in time. The optimization program may report the progress by printing and/or plotting the history data of knob variables and objective function values. Algorithm behaviors can also be reported.

A graphic user interface (GUI) can be used to set up the optimization problem. The progress data can be printed and plotted on the GUI. It is also possible to provide the ability for interrupting and resuming the algorithm.

7.2 REVIEW OF OPTIMIZATION ALGORITHMS

Function optimization is an extensively researched area. There are numerous optimization algorithms, which cannot be thoroughly reviewed in the scope of this book. Here we only intend to discuss some well known algorithms that are potentially applicable to online optimization.

The optimization problem defined in Eq. (7.2) has constraints on the input variables, while many of the classical optimization algorithms are for unconstrained cases. However, the convergence path from the initial solution to the optimum is often not affected by the parameter boundary. The parameter ranges may serve as a sanity check, as well as a safety insurance. The simple constraints on the parameter ranges in Eq. (7.2) are easy to enforce. For example, when the trial solution is outside the boundary, the objective function is not evaluated on the machine; instead, a large function value is assigned according to its distance from the boundary. Since most optimization algorithms will steer away from areas with large function values, the trial solution will likely move back into the valid parameter space. A not-a-number (NaN) value may also be assigned when the trial solution is outside of the valid parameter range, although in this case the algorithms have to be implemented to handle the NaN value properly. A simpler approach would be for the algorithm to stop when the boundary is reached.

Here we will consider general unconstrained optimization algorithms for multi-variable, nonlinear functions. The traditional optimization algorithms in this area can be classified into two groups, deterministic and stochastic algorithms. The development of machine learning has introduced new techniques to the optimization field, which may be characterized as model-based algorithms. For noise free functions, the convergence path from any initial point is fixed for the deterministic algorithms. On the contrary, the stochastic algorithms have different paths every time as they employ some randomness in the choice of the trial solutions. Model-based algorithms build models with the measurement data and use the models to guide the search for the optimum.

7.2.1 Deterministic optimization algorithms

The deterministic algorithms may be divided into two camps,

- gradient-based methods, and

- gradient-free methods.

7.2.1.1 Gradient-based methods

The gradient-based methods include those that require the calculation of the derivatives of the objective function. The objective function in the vicinity of the present solution can be approximated by

$$f(\mathbf{x}) \approx f(\mathbf{x}_0) + \mathbf{b}^T \Delta \mathbf{x} + \frac{1}{2} \Delta \mathbf{x}^T \mathbf{A} \Delta \mathbf{x}, \tag{7.8}$$

where $\Delta \mathbf{x} = \mathbf{x} - \mathbf{x}_0$, $\mathbf{b} \equiv \nabla f(\mathbf{x})|_{\mathbf{x}=\mathbf{x}_0}$ is the gradient at $\mathbf{x}_0$ and $\mathbf{A}$ is the Hessian matrix, with $A_{ij} = \frac{\partial^2 f}{\partial x_i \partial x_j}|_{\mathbf{x}=\mathbf{x}_0}$. The derivatives provide information of the function distribution around the present solution and hence can be used to guide the search for the minimum.

Gradient-descent: Some methods use only the first order derivatives, i.e. the gradient. One example is the steepest descent algorithm, also known as the gradient descent algorithm. Starting from an initial solution $\mathbf{x}_0$, it looks for the minimum iteratively. At each iteration, it performs a line minimization along the gradient direction, $-\nabla f(\mathbf{x}_i)$, i.e., to find α_i such that $f(\mathbf{x}_i - \alpha_i \nabla f(\mathbf{x}_i))$ is minimized, and makes a step change to the new solution from the present solution $\mathbf{x}_i$,

$$\mathbf{x}_{i+1} = \mathbf{x}_i - \alpha_i \nabla f(\mathbf{x}_i). \tag{7.9}$$

The line minimization needs not to be exact; it suffices to find a step that considerably reduces the objective function and the gradient. The gradient descent method guarantees the convergence toward the local minimum. However, for nonlinear functions, the local gradient direction is often not the shortest direction to the local minimum. Therefore, the method may result in a zig-zag convergence path with many small steps, which could be very inefficient.

Extremum Seeking (ES): Extremum Seeking is a group of adaptive control methods that attempt to optimize the performance of a dynamic system and maintain a steady state on the extremum [113]. The ES methods optimize functions by approximating the gradients through function evaluations, although the gradients are not formally computed.

In an ES scheme that has recently found applications in the accelerator community, the knob parameters are varied from iteration n to iteration $n+1$ by adding an oscillatory term that is modulated by the objective function [108, 107],

$$x_i(n+1) = x_i(n) + \Delta \sqrt{\alpha \omega_i} \cos(\omega_i n \Delta + k f(\mathbf{x}_n)), \tag{7.10}$$

where Δ, α, k, and ω_i, $i = 1-N$, are parameters that control the behavior of the algorithm. ω_i is the rotation rate of parameter x_i. Each knob parameter should have a different ω_i value. Δ is a small number that can be considered as the step size. A typical choice of its value may be $\Delta = \frac{2\pi}{20 \max(\omega_i)}$, such that it takes at least 20 iterations to complete one rotation if the objective function is at the extremum. $k > 0$ is required for a minimization problem and its value should be chosen according to the objective function value and the step size. α controls the rotation amplitude of the knob parameters.

For the knob parameter x_i, the oscillation phase increment in iteration n is $\omega_i \Delta + k \Delta f_n$, where $\Delta f_n = f(\mathbf{x}_{n+1}) - f(\mathbf{x}_n)$. If the phase is in a proper value such that the objective function is being reduced, hence $k\Delta f_n < 0$, the phase rotation will be slowed down and the objective function will continue to decrease. It can be shown that in the limiting case of large ω_i and small Δ, the average behavior of the knob parameters is to follow the direction of gradient descent. If the extremum is reached, all knob parameters will continue to rotate with a constant amplitude and rotation rate, at which point the α parameter can be decreased or set to zero to effectively turn the algorithm off.

One advantage of the ES method is that it can be applied to an objective function that is dynamically varying, in which case the algorithm will automatically chase the drifting extremum. This has been demonstrated in an experiment on the SPEAR3 ring [107]. However, generally the ES method is not very efficient. And for any application problem the algorithm parameters need to be carefully adjusted in order for the method to work. Finding the appropriate algorithm setting for a new application can be time consuming.

Newton method: Some algorithms require the calculation of the second order derivatives, i.e., the Hessian matrix. These include the Newton method, in which the step change from the present solution to the local minimum is calculated using a quadratic approximation of the function, which gives

$$\mathbf{x}_{i+1} = \mathbf{x}_i - \mathbf{A}_i^{-1} \nabla f(\mathbf{x}_i). \tag{7.11}$$

The Newton method converges fast when the starting point is close to the minimum. In the area where the quadratic approximation is not valid, the predicted step change may be unreasonable. In such cases, modifications may be made to Eq. (7.11). For example, the step change may be scaled down by a factor $\lambda_i < 1$, in which case the method is called the relaxed Newton method. Another possible modification is to add a diagonal, positive definite matrix to the Hessian, as is done in the Levenberg-Marquardt method,

$$\mathbf{x}_{i+1} = \mathbf{x}_i - (\mathbf{A}_i + \lambda_i \mathbf{I})^{-1} \nabla f(\mathbf{x}_i), \tag{7.12}$$

where $\mathbf{I}$ is the unit matrix.

Quasi-Newton methods: Quasi-Newton methods use the Hessian matrix but do not require the explicit calculation of it. Instead, an approximate Hessian matrix or its inverse is built up using the previous solutions and the gradients. For example, for the Davidon-Fletcher-Powell (DFP) method, the inverse Hessian matrix at iteration $i + 1$, $\mathbf{H}_{i+1}$, is updated with [97]

$$\mathbf{H}_{i+1} = \mathbf{H}_i + \frac{\Delta\mathbf{x}_{i+1}\Delta\mathbf{x}_{i+1}^T}{\Delta\mathbf{x}_{i+1}^T\Delta\mathbf{g}_{i+1}} - \frac{\mathbf{H}_i\Delta\mathbf{g}_{i+1}\Delta\mathbf{g}_{i+1}^T\mathbf{H}_i}{\Delta\mathbf{g}_{i+1}^T\mathbf{H}_i\Delta\mathbf{g}_{i+1}}, \tag{7.13}$$

where $\Delta\mathbf{x}_{i+1} = \mathbf{x}_{i+1} - \mathbf{x}_i$ and $\Delta\mathbf{g}_{i+1} = \nabla f(\mathbf{x}_{i+1}) - \nabla f(\mathbf{x}_i)$. Another widely used quasi-Newton method is the BFGS algorithm, which uses a slightly different updating formula for $\mathbf{H}$. Eq. (7.11) is used to update the solution, with $\mathbf{A}^{-1}$ replaced with $\mathbf{H}$. The initial inverse Hessian matrix may be set to the identity matrix. It can be shown that for quadratic functions, after some iterations, $\mathbf{H}$ will be a good approximation of the inverse Hessian matrix.

Conjugate-gradient method: The conjugate gradient method achieves a similar performance with the quasi-Newton methods without building up the inverse Hessian matrix. It starts with a line search in the steepest descent direction at the initial solution, $\mathbf{x}_0$, to find parameter α_0 that minimizes $f(\mathbf{x}_0 + \alpha_0 \mathbf{h}_0)$, where $\mathbf{h}_0 = \mathbf{g}_0 = -\nabla f(\mathbf{x}_0)$. It subsequently moves to the solution $\mathbf{x}_1 = \mathbf{x}_0 + \alpha_0 \mathbf{h}_0$. Later at each iteration, it computes the local gradient $\mathbf{g}_i = -\nabla f(\mathbf{x}_i)$, and updates the conjugate direction with $\mathbf{h}_i = \mathbf{g}_i + \beta_i \mathbf{h}_{i-1}$, with β_i given by [97]

$$\beta_i^{\mathrm{FR}} = \frac{\mathbf{g}_i^T \mathbf{g}_i}{\mathbf{g}_{i-1}^T \mathbf{g}_{i-1}} \quad \text{or} \quad \beta_i^{\mathrm{PR}} = \frac{\mathbf{g}_i^T (\mathbf{g}_i - \mathbf{g}_{i-1})}{\mathbf{g}_{i-1}^T \mathbf{g}_{i-1}}, \tag{7.14}$$

where β^{FR} is by the Fletcher-Reeves formula and β^{PR} by the Polak-Ribiere formula. β_i is often set to zero if the calculated value is negative. A line search is then performed to find α_i that minimizes the objective function in the conjugate direction, i.e., to minimize $f(\mathbf{x}_i + \alpha_i \mathbf{h}_i)$, and the solution moves to $\mathbf{x}_{i+1} = \mathbf{x}_i + \alpha_i \mathbf{h}_i$.

The above optimization algorithms require the calculation of the gradients, which typically cannot be done for online optimization since analytic forms of the objective functions are not available. The gradients can be approximated with numeric differences in these algorithms. In such cases, we still consider them gradient-based methods because the same principles are used.

Traditional implementations of the gradient-based methods usually do not work for online applications because they often assume smooth objective functions and hence use a tiny step size in numeric difference calculations, for which the corresponding changes of the function value may be easily overwhelmed by the measurement noise. The step size has to be carefully controlled such that the variation of the actual function value dominates the random noise and in the same time remains in the linear region in order to obtain a valid approximation of the gradient. This may not be easy to do without prior knowledge of the objective function. Errors in the gradients will distort the convergence path and can cause the algorithms to fail. Obviously, errors in the Hessian matrix or the conjugate directions due to measurement noise will be even bigger than errors in the first derivatives and are likely to have big impact over the performance of the algorithms. It would be worthwhile to study the impact of the measurement noise to the gradient-based methods for online optimization and to develop ways of mitigation.

7.2.1.2 Gradient-free methods

Gradient-free deterministic algorithms include direct search methods and other methods [80]. Direct search methods follow pre-specified routines to search the parameter space and in the routines new trial solutions are chosen only according to the ranking (i.e., comparison results) of the previously evaluated function values; the numeric values are not used. Direct search methods include pattern search, Nelder-Mead simplex search [89], and methods with adaptive search directions.

Nelder-Mead simplex method: The Nelder-Mead simplex method, also known as the downhill simplex method, is probably the most well-known direct search method. This method maintains a non-degenerate simplex during the course of the search. A simplex is an $n + 1$ polytope in an n-dimension space. For example, a simplex in the 2-dimensional space is a triangle and a simplex in a 3-dimensional space is a tetrahedron. The algorithm first builds a simplex around the initial solution, typically with it as one vertex and generating the other n vertices by taking a step in one of the n parameters each. During an iteration, it performs the following operations

1. Sorts the function values on the vertices. Label the vertices 1 through $n + 1$ with the corresponding function values in the ascending order, i.e., $f_1 \leq f_2 \leq \cdots \leq f_{n+1}$, where $f_i \equiv f(\mathbf{x}_i)$ and $\mathbf{x}_i$ is the i'th vertex. Calculate the center point of the simplex face opposite to vertex $\mathbf{x}_{n+1}$ (which has the largest function value), $\mathbf{x}_c \equiv \frac{1}{n} \sum_{i=1}^{n} \mathbf{x}_i$. On the line connecting $\mathbf{x}_{n+1}$ and $\mathbf{x}_c$, the following points are defined:

   Reflection point: $\mathbf{x}_r = \mathbf{x}_c + (\mathbf{x}_c - \mathbf{x}_{n+1})$.

   Expansion point: $\mathbf{x}_e = \mathbf{x}_c + 2(\mathbf{x}_c - \mathbf{x}_{n+1})$.

   Inner contraction: $\mathbf{x}_{ic} = \mathbf{x}_c - \frac{1}{2}(\mathbf{x}_c - \mathbf{x}_{n+1})$.

   Outer contraction: $\mathbf{x}_{oc} = \mathbf{x}_c + \frac{1}{2}(\mathbf{x}_c - \mathbf{x}_{n+1})$.

2. *Reflection:* Evaluate the function value at the reflection point, $f_r = f(\mathbf{x}_r)$. If the value at the reflection point is better than the second worst, but not better than the best vertex, i.e., $f_1 \leq f_r < f_n$, replace vertex $\mathbf{x}_{n+1}$ with $\mathbf{x}_r$ and finish the iteration.

3. *Expansion:* If the reflection point is better than the best vertex, i.e., $f_r < f_1$, evaluate the expansion point. Replace $\mathbf{x}_{n+1}$ with the better of the reflection point and the expansion point and finish the iteration.

4. *Outer contraction:* If $f_n < f_r < f_{n+1}$, evaluate the outer contraction point. If $f_{oc} = f(\mathbf{x}_{oc}) < f_r$, replace vertex $\mathbf{x}_{n+1}$ with $\mathbf{x}_{oc}$ and finish the iteration.

5. *Inner contraction:* If $f_r > f_{n+1}$, evaluate the inner contraction point. If $f_{ic} = f(\mathbf{x}_{ic}) < f_r$, replace vertex $\mathbf{x}_{n+1}$ with $\mathbf{x}_{ic}$ and finish the iteration.

6. *Shrink:* If none of the above operations terminates the iteration, shrink the size of the simplex toward the best vertex, $\mathbf{x}_1$, by replacing all other vertices with $\mathbf{x}_i' = \mathbf{x}_1 + \frac{1}{2}(\mathbf{x}_i - \mathbf{x}_1)$, with $i = 2, 3, \cdots, n + 1$. Evaluate the function values on all new vertices and go on to the next iteration.

The operations of the downhill simplex method are illustrated in Figure 7.2 for the 2-dimensional case. In the above procedure, in each iteration the worst

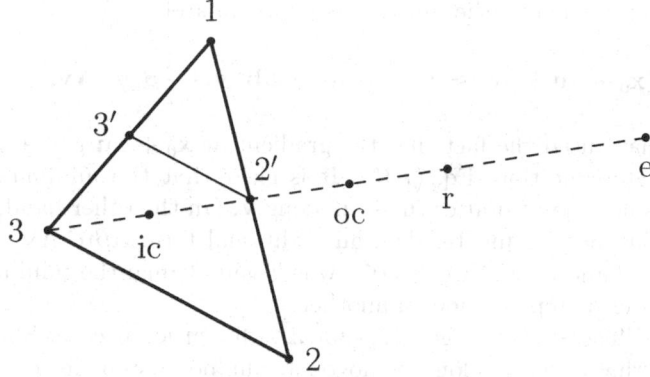

Figure 7.2 Illustration of the downhill simplex method in the 2-dimensional case. Function values on the vertices are ordered $f_1 \leq f_2 \leq f_3$.

vertex is replaced with a new point, which is selected according to a series of trial and comparison operations. The Nelder-Mead simplex method works in most times and it is very efficient when it does. However, there are cases when the method fails, with symptoms like slow convergence or premature convergence to non-optimal points.

Direction search algorithms tend to be sensitive to the noise in the function evaluations as they completely rely on the comparison results of function values to make decisions on the search paths. The noise will likely change the comparison results and hence in turn change the convergence path or cause the algorithm to randomly wander around without converging.

Besides the direct search algorithms, other gradient-free deterministic algorithms use the numeric function values to guide the selection of new trial solutions. This is done, for example, by modeling the objective function within the neighborhood of the best solution. One such method is Powell's conjugate direction method [96].

Powell's method: Powell's method performs iterative one dimensional optimization over a set of directions that are linearly independent and mutually conjugate. In the parameter space a direction is represented by a unit vector. Two directions, $\mathbf{u}$ and $\mathbf{v}$, are mutually conjugate for the optimization problem if they satisfy

$$\mathbf{u}^T \mathbf{A} \mathbf{v} = 0, \tag{7.15}$$

where $\mathbf{A}$ is the Hessian matrix. The benefit of searching along the conjugate directions is that a move along one direction does not change the position of the best solution in the other directions (to the extent that the quadratic approximation of the objective function is valid). Suppose a step α is taken to minimize the objective function along the $\mathbf{u}$ direction, after that a second step β along $\mathbf{v}$ is made, the function is approximately

$$f(\mathbf{x}_0 + \alpha\mathbf{u} + \beta\mathbf{v}) \approx f(\mathbf{x}_0 + \alpha\mathbf{u}) + \beta\mathbf{b}^T\mathbf{v} + \frac{1}{2}\beta^2\mathbf{v}^T\mathbf{A}\mathbf{v}, \qquad (7.16)$$

where we have used the fact that the gradient at $\mathbf{x}_0 + \alpha\mathbf{u}$ is $\mathbf{b} + \alpha\mathbf{A} \cdot \mathbf{u}$ and the conjugate condition, Eq. (7.15). It is clear that the minimum in the $\mathbf{u}$ direction is not changed after the step along $\mathbf{v}$. On the other hand, if the two directions are not conjugate, then an additional term $\alpha\beta\mathbf{u}^T\mathbf{A}\mathbf{v}$ will appear on the right hand side of Eq. (7.16), which will change the minimum in one direction after a step is made on another.

This is illustrated in Figure 7.3 for a 2-dimensional case. Starting from point 0, if the search is along orthogonal, but non-conjugate directions, for example, $\mathbf{x}_1$ and $\mathbf{x}_2$, the convergence path will consist of many small segments in the corridor leading to the minimum. However, if the search is along the conjugate directions, $\mathbf{u}_1$ and $\mathbf{u}_2$, the minimum will be found in only two steps.

If the Hessian matrix can be calculated and is positive definite, the conjugate directions can be determined from its eigenvectors. If the conjugate directions cannot be calculated, Powell's method can construct the conjugate direction set from the successive line minimizations with a non-degenerate initial direction set. The initial directions may be simply the unit vectors along the parameter axes. Suppose at the beginning of an iteration, the solution is labeled $\mathbf{x}_0$ and $f_0 = f(\mathbf{x}_0)$, and the directions are $\mathbf{u}_k$, $k = 1, 2, \cdots, n$. During the iteration, the algorithm executes the following steps [97],

1. Perform line minimizations along all n directions and record the biggest function value drop during one line minimization, call it Δ and mark the corresponding direction $\mathbf{u}_d$. The final new solution is labeled $\mathbf{x}_m$, with $f_m = f(\mathbf{x}_m)$.

2. Evaluate the function value at the extension point, $\mathbf{x}_t = 2\mathbf{x}_m - \mathbf{x}_0$, with $f_t = f(\mathbf{x}_t)$.

3. If $f_t > f_0$ or $2(f_0 + f_t - 2f_m)(f_0 - f_m - \Delta)^2 > \Delta(f_0 - f_t)^2$, terminate the iteration without replacing a direction; otherwise, replace the direction $\mathbf{u}_d$ with the new direction $\mathbf{u}_m = \frac{\mathbf{x}_m - \mathbf{x}_0}{||\mathbf{x}_m - \mathbf{x}_0||}$. Perform the line minimization along $\mathbf{u}_m$ and terminate the iteration.

Replacing a direction with the direction that goes from the initial solution to the final solution is desired because the latter is likely a more efficient search direction. For example, in Figure 7.3, the direction from point 0 to 2 is much better aligned with the corridor leading to the local minimum. The direction with the largest function value drop is chosen to be replaced because it is likely to have a large overlap with the new direction; removing it from the direction set reduces the likelihood of introducing degeneracy. In Step 2 the extension point is evaluated as a way to check the validity of the new direction. If either of the two conditions in Step 3 is met, there is not much to gain in this direction and the replacement is skipped. In general, for a quadratic function, a conjugate direction set will be obtained after n iterations.

A critical component of Powell's method is the line minimization, which performs the task of minimizing the 1-dimensional problem of $g(\alpha) = f(\mathbf{x}_0 + \alpha\mathbf{u})$. One of two derivative-free line optimizers, the golden section search method and Brent's inverse parabolic interpolation method, is often used. For both line optimizers, the local minimum is first bracketed in a zone $\alpha \in [a, b]$. This is ensured if a third point c is found between points a and b for which $g(c) < g(a)$ and $g(c) < g(b)$ are both satisfied.

With the initial three points, $a < c < b$, the golden section search then samples a new point t in one of the two subdivisions, say $t \in (a, c)$. If $f(t) < f(c)$, then $[a, c]$ becomes the new bracket; if $f(t) > f(c)$, $[t, b]$ becomes the new bracket. It can iteratively proceed until the minimum is found with the desired tolerance. For the highest efficiency, the distance from the inside sample point to the bracket boundary is chosen to be $\frac{\sqrt{5}-1}{2} \approx 0.618$. For Brent's method, the function values at point a, c, and b are used to construct a parabola and the location corresponding to the minimum of the parabola is used as the new sample point. The golden section method has linear convergence, while the inverse parabola interpolation has quadratic convergence. The latter is used more often for smooth functions. With noise in the function values, both the golden section search and inverse parabola interpolation methods could fail.

When the golden section search method is used, Powell's method can still be characterized as a direct search method. However, with the inverse parabolic interpolation, it no longer qualifies as a direct search method since the function values are used to construct a model. Powell's method is especially efficient for convex functions.

The Nelder-Mead simplex method and Powell's method are both popular and powerful optimization algorithms for smooth functions. Noise in the function values has a big impact to the performance of both methods. Modifications to the original algorithms can be made to mitigate the impact of noise. This will be discussed in Section 7.3.

7.2.2 Stochastic optimization algorithms

Some optimization algorithms make use of random operations in the course of searching for the optimal solutions. The random operations could be choosing the parameter values of the trial solutions or making a decision of accepting or rejecting a solution. Because of the randomness, the convergence path is different every time. These algorithms are called stochastic optimization algorithms.

Stochastic optimization algorithms often have poorer efficiency than the deterministic methods in terms of the number of required function evaluations to converge to the optimum. This is understandable as these methods are not "greedy": they do not intend to take the shortest paths toward the minimum. As the new trial solutions are somewhat randomly chosen, the outcome of

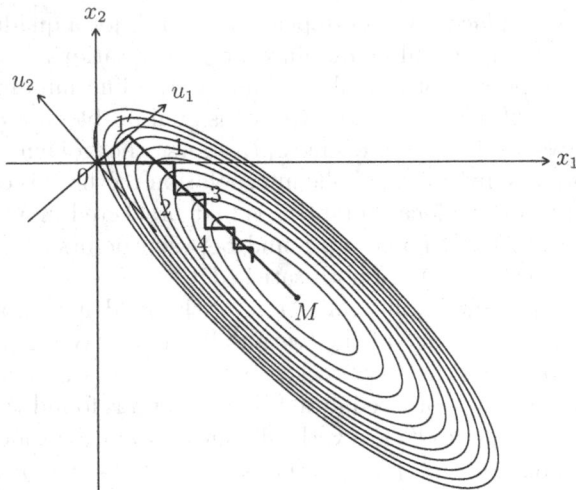

Figure 7.3 Illustration of iterative line minimization with or without conjugate directions. Starting from the initial solution at point 0, the algorithms try to converge to the minimum at point M with non-conjugate directions $\mathbf{x}_1$ and $\mathbf{x}_2$, or the conjugate directions, $\mathbf{u}_1$ and $\mathbf{u}_2$.

function evaluations is not always an improvement. However, there is a benefit at the cost of the efficiency loss – the stochastic algorithms often have a better chance of finding the global optimum. By allowing taking steps in bad directions and using random sampling, the stochastic algorithms are not as easily attracted to the local minima as the deterministic algorithms are.

Random search: One of the simplest stochastic optimization algorithms is to sample the parameter space with randomly selected solutions. This method is not efficient, but it can be very useful in some cases. For example, it can be used to find working solutions from which a starting point for other algorithms is chosen.

Random search [99] is a method with a little more sophistication. It searches in the vicinity of the present solution with a random trial solution, e.g.,

$$\mathbf{x}_{i+1} = \mathbf{x}_i + \Delta\mathbf{x}, \quad \text{with } ||\Delta\mathbf{x}|| \le r, \tag{7.17}$$

where $\Delta\mathbf{x}$ is randomly selected within the hyper-sphere with a radius r. The algorithm moves to the new solution if it is better, otherwise tries a new one.

Simulated annealing [97]: Simulated annealing is a method that mimics the slow cooling process through which hot materials settle to the state with the lowest energy. The system is characterized by its energy E and temperature T. The system energy normally tends to decrease; but it can also increase with a certain probability. The probability for the system energy to change from E_0 to E is given by $p(E, E_0, T) = \exp(-\frac{E-E_0}{kT})$ for $E > E_0$ and $p = 1$ for

$E \leq E_0$. When kT is large, there is a large probability for the system energy to increase. This allows the system to jump out of local minima. However, when temperature is cooled down, the system energy will more likely decrease.

When simulated annealing is applied to function minimization, the function value is analogous to the system energy. The temperature is a controlled parameter which dictates the probability distribution of function value increase, $p(f, f_0, T)$. In each step, a new temperature T is chosen. A trial solution around the present solution is evaluated. The new solution is accepted if its function value is lower (i.e., $f < f_0$). If the function value is higher, the new solution is accepted if $p(f, f_0, T)$ is larger than a random number drawn from the uniform distribution of $[0, 1]$. The variation of temperature with the steps, the way to choose the trial solution, and the choice of the probability function have a large impact to the performance of the algorithm.

Genetic algorithms (GA) [28, 27]: Genetic algorithms became popular in accelerator design optimization in recent years. A genetic algorithm manipulates a population of solutions over many generations. In each generation, a portion of the population is replaced with good solutions selected from new solutions that are generated through cross-over or mutation operations. In the cross-over operation two "children" solutions are spawned by combining the parameter values of two "parent" solutions. The mutation operation generates a new solution by randomly modifying the parameter values of an existing solution. The new solutions are first mixed with the existing population of solutions, from which the fittest solutions (i.e., the ones with the lowest objective function values) are selected to enter the next generation. The solutions that survive the selection operation are generally better and they tend to produce better new solutions. Therefore, the fitness (i.e., the objective function) of the solutions will improve over time and the population gradually converges to the minimum. The NSGA-II algorithm is a popular multi-objective genetic algorithm (MOGA) [27].

The parameter vector for a solution is called a chromosome, which can be represented by a bit string or an array of floating numbers. For a bit string, the cross-over can be done by swapping bits between the two chromosomes. In the case of an array of floating numbers, cross-over of two solutions can be performed with simulated binary cross-over (SBX) [29],

$$x'_{1,k} = \frac{1}{2}[(1 - \beta_k)x_{1,k} + (1 + \beta_k)x_{2,k}], \tag{7.18a}$$

$$x'_{2,k} = \frac{1}{2}[(1 + \beta_k)x_{1,k} + (1 - \beta_k)x_{2,k}], \tag{7.18b}$$

where $\mathbf{x}_1$ and $\mathbf{x}_2$ are the parent solutions, $\mathbf{x}'_1$ and $\mathbf{x}'_2$ are the children solutions, subscript k indicates the k'th parameter, and β_k is a random number given by

$$\beta(u \leq \frac{1}{2}) = (2u)^{\frac{1}{\mu_c+1}}, \text{ or } \beta(u > \frac{1}{2}) = (2(1 - u))^{-\frac{1}{\mu_c+1}}, \tag{7.19}$$

where u is a random number drawn from the uniform distribution in $(0,1)$ and μ_c is a control parameter.

The mutation operation is performed by adding a random variation to each parameter,

$$x'_k = x_k + L_k \delta_k, \tag{7.20}$$

where L_k is the range of the x_k parameter and δ_k is a random number calculated with

$$\delta(v < \frac{1}{2}) = (2v)^{\frac{1}{\mu_m+1}} - 1, \text{ or } \delta(v > \frac{1}{2}) = 1 - (2(1-v))^{\frac{1}{\mu_m+1}}, \tag{7.21}$$

with random variable v drawn from the uniform distribution in $(0,1)$ and μ_m is another control parameter.

For problems with a single objective function, the selection of solutions to enter the next generation is done by simple sorting. Genetic algorithms can be conveniently extended to multi-objective problems by applying non-dominated sorting to select the fittest solutions. Solutions in the leading fronts enter the next generation until the quota is filled up. When the last front with qualified solutions yields more solutions than needed, the solutions can be randomly picked or chosen to maximize the diversity of the surviving population, using the so-called crowding distance as the criterion. Constraints are easy to implement for genetic algorithms as they can be considered as a part of the fitness criteria; solutions that violate the constraints are given lower fitness.

The initial population can be generated randomly throughout the parameter space. However, it often works better if it is randomly generated in a small neighborhood around a good solution.

Genetic algorithms are powerful methods for design optimization, yet they also have limitations. The algorithms may have low efficiency as many of the children solutions are similar to the parent solutions that have been previously evaluated. It may be difficult to apply the genetic algorithms for high dimension problems as it would require a large population of solutions for the algorithms to work and the corresponding computational cost for evaluating these solutions could be prohibitive. The algorithms can converge prematurely to a non-optimal region when all surviving solutions are similar. Improving the percentage of mutation can alleviate the problem of premature convergence, but it will also slow down the speed of convergence.

When used for the online optimization, genetic algorithms suffer an additional limitation. Negative values in the measurement noise will give some solutions an advantage over the others. These solutions tend to survive the selection operation, driving out the real good solutions and thus defeat the working principles of the algorithm. Re-sampling for the population can mitigate the noise problem [14].

Particle swarm optimization (PSO) [71, 92, 59]: Particle swarm optimization is similar to the genetic algorithms in that it also manipulates a population of solutions. Here the solutions are considered as moving "particles" in the parameter space. The coordinates of the particles are the parameter values. The coordinates of each particle change in every iteration by an increment called its "speed",

$$\mathbf{x}_i^{k+1} = \mathbf{x}_i^k + \mathbf{v}_i^k, \tag{7.22}$$

where $\mathbf{v}_i^k$ is the speed for particle i at iteration k. The speed consists of contributions from three terms,

$$\mathbf{v}_i^{k+1} = w\mathbf{v}_i^k + c_1 r_1 (\mathbf{p}_i^k - \mathbf{x}_i^k) + c_2 r_2 (\mathbf{g}_i^k - \mathbf{x}_i^k), \tag{7.23}$$

where on the right hand side the first term represents the previous speed, the second term represents the acceleration toward the best solution on the trajectory (denoted by $\mathbf{p}_i^k$) traversed by the particle, and the third term represents the acceleration toward the overall best solution (denoted by $\mathbf{g}_i^k$). r_1 and r_2 are random numbers drawn from the zone $[0, 1]$. The best solutions on the trajectories and the overall best solution are updated in each iteration. In case of a multi-objective optimization, a population of global best solutions is kept and a randomly selected one is used for $\mathbf{g}_i^k$ every time. The control parameters w, c_1, and c_2 can be changed to adjust the algorithm behavior. The values of $w = 0.4$ and $c_1 = c_2 = 1$ can be used. A small percentage of the new particle coordinates can also be generated from the previous coordinates with a mutation operation as is done in genetic algorithms.

The initial swarm of particles can be generated randomly around a known good solution or throughout the parameter space. The initial speed is also randomly specified. The magnitude of the initial speed may be chosen to be a small fraction, e.g., 10%, of the parameter ranges. After the initial swarm of particles is launched, it will sweep through the parameter space and be attracted toward the optima.

Particle swarm optimization is also suitable for multi-objective problems. One only needs to use non-dominated sorting in the selection of personal and global best solutions.

Experiences of applying PSO and GA algorithms on the same problems indicate that the PSO method is often more efficient than the GA method [92, 59]. It converges faster because it has high diversity in the new solutions and hence there is less waste of time spent on re-evaluating known solutions.

7.2.3 Machine learning (ML) methods

Machine learning is the collection of advanced computer algorithms that can analyze sample data to extract patterns or build models, which in turn are to be used to make predictions or decisions with new data. Machine learning techniques can be classified into three categories: supervised learning, unsupervised learning, and reinforcement learning. These techniques can be used in function optimization by modeling the parameter space and providing effective strategies in exploring the parameter space.

Some traditional optimization algorithms employ simple modeling of the sample data. For example, the derivatives or parabolic fitting can be considered local models around one point. But these algorithms do not attempt to build models in an extended area of the parameters space. There are algorithms that take all or a significant subset of previously evaluated data points into account in order to build a model that represents the objective function, which is in turn used to guide the search for the optimum. These algorithms are considered machine learning methods as they actually learn about the objective function through probing the parameter space.

Gaussian Process (GP) optimizer: Gaussian Process optimization is a type of Bayesian optimization [72, 129, 87, 69, 98]. Bayesian optimization combines prior assumptions and the observed data to establish a posterior model of the objective function, based on Bayes' theorem of conditional probabilities. Specifically, given the prior probability distribution of the objective, $P(f)$, and the likelihood of measuring the sample data under function f, the posterior function can be obtained with

$$P(f|\mathcal{D}_t) = P(\mathcal{D}_t|f)P(f), \tag{7.24}$$

where $P(a|b)$ denotes the conditional probability of a under condition b, $\mathcal{D}_t = \{(\mathbf{x}_i, f(\mathbf{x}_i)|i \in (1, 2, \cdots t)\}$, and t is the number of data points. $P(f|\mathcal{D}_t)$ is a surrogate function that approximates the actual objective, which can be used to help find the optimum, for example, by predicting the next trial solution that has the best chance to minimize the objective.

In a GP optimizer the prior and posterior distributions of the objective function are both Gaussian processes. A Gaussian process is a multi-variate normal distribution of variables distributed over space or time. The function values over all points in the parameter space is a Gaussian process, which is characterized by the mean value function, $m(\mathbf{x})$, and the covariance function between any pair of points, $k(\mathbf{x}, \mathbf{x}')$. The prior mean function is usually assumed $m(\mathbf{x}) = 0$. The covariance, known as the kernel function, is commonly chosen to be the squared exponential function,

$$k(\mathbf{x}, \mathbf{x}') = \Sigma_f^2 \exp\left(-\frac{1}{2}(\mathbf{x} - \mathbf{x}')^T \Theta^{-2}(\mathbf{x} - \mathbf{x}')\right), \tag{7.25}$$

where Σ_f^2 is the prior variance of f over the parameter space, $\Theta = \text{diag}(\theta_1, \theta_2, \cdots, \theta_n)$, and θ_i is a parameter that characterizes the correlation length of the objective function over parameter x_i. A large θ_i indicates that the function values at two points separated with a large distance in the x_i parameter are still highly correlated.

After the data sample $\mathcal{D}_t$ is obtained, we would like to know the posterior distribution of function value $f_{t+1} = f(\mathbf{x}_{t+1})$ at an arbitrary trial solution $\mathbf{x}_{t+1}$. This can be derived from the joint distribution of f_{t+1} and the function values on the previous data points, $\mathbf{f}_{t+1} = (\mathbf{f}_t, f_{t+1})$, with $\mathbf{f}_t = (f_1, f_2, \cdots, f_t)$.

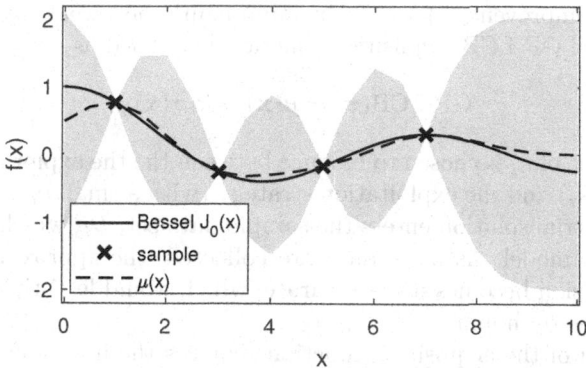

Figure 7.4 The Gaussian process model for the Bessel $J_0(x)$ function with four data points. Shaded area indicates the $\pm 2\sigma$ confidence.

The prior joint distribution of $\mathbf{f}_{t+1}$ is a normal distribution given by

$$\mathcal{N}\left(\mathbf{0}, \begin{pmatrix} \mathbf{K} & \mathbf{k} \\ \mathbf{k}^T & k(\mathbf{x}_{t+1}, \mathbf{x}_{t+1}), \end{pmatrix}\right), \tag{7.26}$$

where the kernel matrix, $\mathbf{K}$, is a $t \times t$ matrix with elements, $K_{ij} = k(\mathbf{x}_i, \mathbf{x}_j)$, and $\mathbf{k}$ is a $t \times 1$ row vector whose elements are $k_i = k(\mathbf{x}_i, \mathbf{x}_{t+1})$. From the joint distribution and the evidence of the measured data set, $\mathcal{D}_t$, the posterior distribution of f_{t+1} (the conditional distribution with the given data set) is found to be a normal distribution [98],

$$P(f|\mathcal{D}_t) = \mathcal{N}(\mu_{t+1}, \sigma_{t+1}^2), \tag{7.27}$$

where

$$\mu_{t+1} = \mathbf{k}^T \mathbf{K}^{-1} \mathbf{f}_t, \quad \sigma_{t+1}^2 = k(\mathbf{x}_{t+1}, \mathbf{x}_{t+1}) - \mathbf{k}^T \mathbf{K}^{-1} \mathbf{k}. \tag{7.28}$$

The expected mean value, μ_{t+1}, is an approximation of the function f, while the standard deviation σ_{t+1} gives an estimate of the uncertainty, for any point $\mathbf{x}_{t+1}$ throughout the parameter space.

Figure 7.4 shows an example of approximating a 1-dimensional function with a Gaussian process. Four data points sampled from the Bessel $J_0(x)$ function are used to build a posterior model, with $\theta = 1$ assumed for the prior. The true function, the sample points, and the expected mean value are plotted. The shaded area shows the 2σ confidence region for the Gaussian process.

In GP optimization, the posterior model is used to choose a new trial solution by optimizing the acquisition function. There are multiple ways to define the acquisition function, for example, by maximizing the probability of improving from the best evaluated solution (PI, probability of improvement) [86], or by maximizing the expected amount of improvement (EI, expected improvement) [82], or by minimizing the lower confidence bound (LCB) [4]. The GP-LCB acquisition function is defined as

$$\text{GP-LCB}(\mathbf{x}) = \mu(\mathbf{x}) - \kappa_t \sigma(\mathbf{x}), \tag{7.29}$$

where the value of κ_t is chosen to balance between the the exploration strategy (with a large κ_t) and the exploitation strategy (with a small κ_t). After evaluation, the new trial solution enters the sample data set, $\mathcal{D}_t$, which is then used to update the model. As more data are collected, the approximation of the objective function becomes more accurate, which would lead the algorithm to converge to the optimum.

Evaluation of the acquisition function requires the inversion of the kernel matrix. Subsequent calculations with the Gaussian process model involve matrix multiplications. As the number of data points increases, the computation time will also increase. This is acceptable if the number of data points is on the order of hundreds. But the computation would become too slow if the data sample size is much larger.

When noise is present in the function evaluations, the diagonal elements of the kernel matrix change as the random noise enters the variance, but the off-diagonal elements do not change (as the noise at different sample point is independent), hence in Eq. (7.28) the kernel matrix is replaced with,

$$\mathbf{K} \to \mathbf{K} + \sigma^2 \mathbf{I}. \tag{7.30}$$

Noise affects the accuracy and even the validity of the Gaussian process model and in turn the performance of the optimizer.

Multi-generation Gaussian process optimizer (MG-GPO): The ability of predicting function values with the posterior GP model can substantially enhance the efficiency of the stochastic optimization algorithms. The MG-GPO [56] algorithm is a method that utilizes this ability to select trial solutions with high potential for the actual function evaluation. It operates iteratively and maintains a fixed number of good solutions, in the same manner as the GA and PSO algorithms. In the case of a multi-objective optimization, a GP model is built for each objective function. Many new solutions are generated for each good solution using cross-over and mutation operations. These solutions are tested with the posterior GP models and are ranked with non-dominated sorting according to the corresponding acquisition function values. Only a fixed number of solutions in the leading fronts are evaluated on the real system. The evaluated solutions are then combined with the existing good solutions, from which a new population of good solutions are selected for the next generation.

The GP models are rebuilt at each generation using only the good solution population and the recently evaluated solutions. Therefore, the size of the sample data set does not grow indefinitely. In simulation, it was shown that the GP-GPO method outperforms both the NSGA-II and MOPSO methods.

7.3 ALGORITHMS FOR ONLINE OPTIMIZATION

An algorithm that is efficient for smooth functions may be unsuitable for online applications if it is sensitive to noise. In this section we first perform tests for some of the algorithms by adding noise to analytic objective functions.

Modifications of two of the popular derivative-free algorithms are described. The robust conjugate direction search (RCDS) method [57] is modified from Powell's method by replacing its 1-dimensional optimizer with a robust optimizer that is aware of the noise level in the functions. The robust simplex (RSimplex) [55] is a modification of the Nelder-Mead simplex method. With the modifications, the algorithms become more tolerant of noise.

7.3.1 Testing of traditional algorithms

A simple analytic function is used to test the performance of the traditional optimization algorithms under noise. The test function has 4 variables and is in the form of

$$f_1(\mathbf{x}) = 1 - e^{-(x_3+x_4)^2} J_0(2\sqrt{r}) + 2(x_3 - 0.5)^2, \text{ with} \qquad (7.31)$$
$$r = 0.5(x_1 + x_2 - 1)^2 + 2(x_1 - x_2)^2 + 0.1(x_1 - x_4 - 1)^2.$$

The parameter ranges are $x_i \in [-2, 2]$ for all 4 variables ($i = 1\text{-}4$). The initial solution is chosen to be $\mathbf{x}_0 = (-0.5, 0, -0.5, 0)^T$. Function $f_1(x)$ has one minimum located at $\mathbf{x}_m = (0.5, 0.5, 0.5, -0.5)^T$, with $y_m = f_1(\mathbf{x}_m) = 0$.

To simulate the effect of measurement noise in online optimization, the function values passed on to the optimization algorithms are modified by adding a Gaussian random variable, ξ, whose standard deviation (referred to as the noise sigma) is σ. Three levels of noise sigmas are used, with $\sigma = 0.001$, 0.01, and 0.1.

Test of the gradient descent method:

We implemented the gradient descent method with an adaptive step length, α, which is increased when the objective function is reduced in a step, and conversely, decreased if the new solution is out of the parameter range or if the function value actually increases. With this we can avoid performing a 1-dimensional optimization for each direction as it will come up with a suitable step length in a few steps, despite the potentially substantial difference in the magnitude of the gradient at different locations and between different objective functions. The initial value of $\alpha = 0.001$ is used. The gradient is calculated with numeric differential. The step size for numerical differentiation in normalized coordinates is chosen to be $\Delta = 0.001$.

In Figure 7.5 the left plot shows the history of the function values for the evaluated solutions during the course of the optimization run for the three noise levels. The number of evaluations is set to 300. Because of the random noise, the convergence path is different every time. The right plot shows the minimum function value reached over the course of 300 evaluations for 100 runs. The curves shown in the left plot correspond to the case for which the

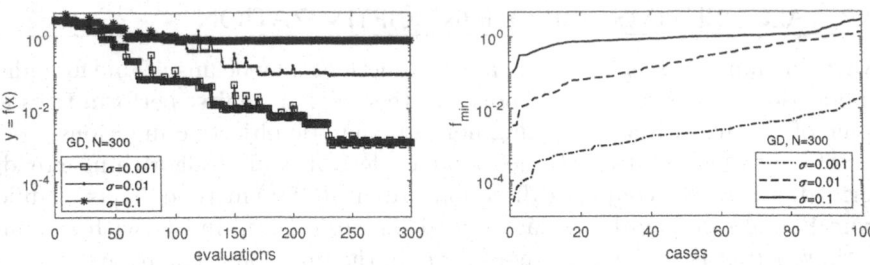

Figure 7.5 Testing of the gradient descent method with the analytic function, $f_1(\mathbf{x})$, defined in Eq. (7.31). Gaussian random noise is added to the function evaluation during optimization, but is not included in the plots. The left plot shows the convergence histories for the typical cases of three different noise levels. The right plot shows the minimum function values for 100 runs in each noise level.

final minimum ranks the 30th for each noise level. The noise term is not included in the function values in both plots.

The gradient descent method works well for the test function when the noise level is low. However, for a high noise level such as $\sigma = 0.01$ or 0.1, it fails to converge to the minimum. A relative large step size ($\Delta = 0.001$) in numerical differentiation is important in this test. If the step size is reduced to $\Delta = 0.0001$, the performance for the $\sigma = 0.01$ cases will be similar to the higher noise cases shown in the figure.

A proper differentiation step size would be important in the application of the gradient descent method to real online optimization problems. The step size may be chosen by requiring the ratio of the noise sigma to the function value change to be less than a certain level, such as 10%. It may not always be possible to satisfy such a condition, though. For example, in the vicinity of the minimum, the slope decreases toward zero.

In the presence of noise, it would be more difficult to properly compute the second order derivatives with numerical differentiation. Therefore, the Newton method or quasi-Newton methods are not considered in the following.

Test of Nelder-Mead simplex method:

The Nelder-Mead simplex method is tested with the same analytic function. The initial simplex is built around the same initial solution, with the initial side lengths set to 0.02 (in normalized coordinates). Figure 7.6 shows the performance for the three noise levels in the same manner as in Figure 7.5, with the left plot for the convergence histories of three typical cases (whose final minimum ranks the 30th) and the right plot for the final minimum for 100 runs. Within 300 evaluations, the algorithm has come to a stop and is no longer making gains in reducing the function value.

With a low noise level ($\sigma = 0.001$), the algorithm successfully converges to the minimum for about 80% of all cases. However, it fails for the other cases. With a higher noise level, $\sigma = 0.01$, it has poorer performance. At the

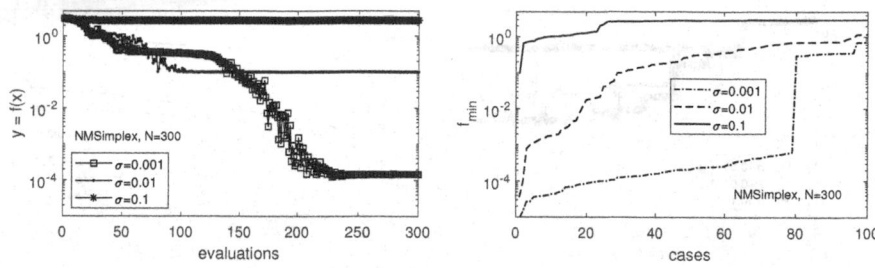

Figure 7.6 Testing of the Nelder-Mead simplex method with function $f_1(\mathbf{x})$ and three random noise levels. Noise is added in function evaluations during optimization, but is not included in these plots. The initial simplex side length is 0.02 in normalized coordinates.

level of $\sigma = 0.1$, the N-M simplex method generally cannot find the minimum. A closer examination shows that when the noise level is comparable to the differences between the function values on the vertices of the simplex, the algorithm starts to deviate from the ideal convergence path. Increasing the initial size of the simplex would help alleviate the issue as it delays the time when that happens.

Test of Powell's method:

Function $f_1(\mathbf{x})$ in Eq. (7.31) is also used to test Powell's method. The original direction set consists of simply the directions along the parameter axes. Figure 7.7 shows the test results. The golden section method is used as the 1-dimensional optimizer for the results shown. Although the golden section method converges more slowly than the Brent's method for smooth functions, it performs better than the latter when there is noise in the function evaluations. This is understandable as the noise could render the quadratic fits invalid and hence lead to wrong predictions. With a relatively low noise level, Powell's method converges to the minimum. However, for the high noise level case ($\sigma = 0.1$), it cannot find the minimum in most cases.

Test of Gaussian Process optimizer:

A GP optimizer has been implemented and tested with the test function in Eq. (7.31). The initial data points used to train the GP model are the vertices of a simplex with side length equal to 0.02 (in normalized coordinates). The acquisition function is chosen to be the GP-LCB with $\kappa_t = 1.8$. The Nelder-Mead simplex method is used to search for the minimum for the acquisition function at each step. The search starts from a random point in the vicinity of the solution with the current minimum.

The kernel function used for the GP is the squared exponential, Eq. (7.25), with $\Sigma = 1.0$ and $\theta_i = 0.1$ for $i = 1$-4. The maximum number of function evaluations is set to 300 for each optimization run. Figure 7.8 shows the test results for the three noise levels. With medium or low noise levels ($\sigma = 0.001$

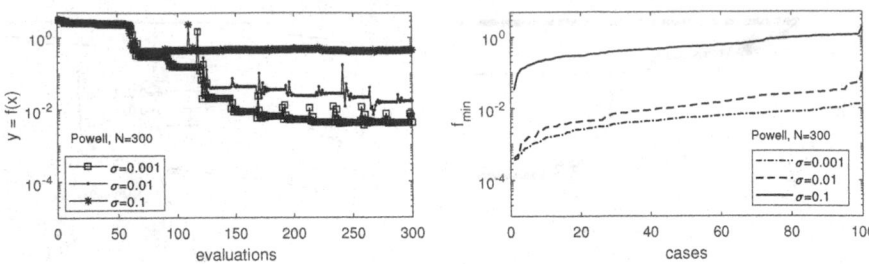

Figure 7.7 Testing of Powell's method with function $f_1(\mathbf{x})$ in Eq. (7.31) and three random noise levels. The golden section method is used as the line optimizer. Left: typical convergence histories (for cases whose final minimum ranks the 30th); right: minimum function values within 300 evaluations for 100 runs.

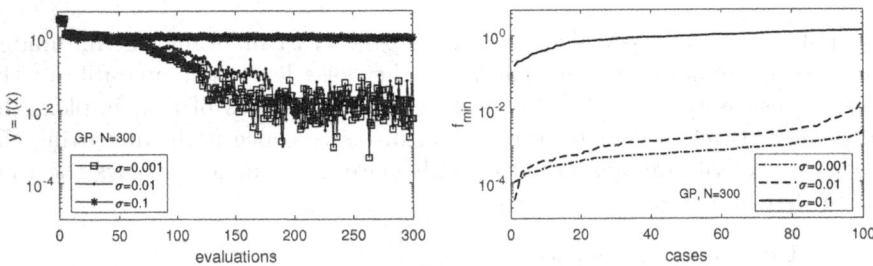

Figure 7.8 Testing of the Gaussian Process optimizer with function $f_1(\mathbf{x})$ and three random noise levels. Left: typical convergence histories; right: minimum function values within 300 evaluations.

and 0.01), the GP optimizer can successfully find the minimum. However, with a high noise level ($\sigma = 0.1$), it fails to converge to the minimum.

It is worth noting that the GP optimizer becomes slow as the number of data points grows. For example, it takes 9 hrs to complete 100 runs for the test while on the same computer it takes only a few seconds for all other algorithms. It would be even slower if the number of function evaluations is larger.

Summary of test results:
From the test results shown in Figures 7.5 – 7.8, it is clear that noise can significantly impact the behaviors of the optimization algorithms. At low noise levels, the algorithms may have similar performances as for smooth functions. However, noise at high levels usually will disrupt the actions of the algorithms and cause failures to converge to the minimum. The level of noise that leads to failures depends on the working principles of the algorithms as well as the nature of the objective functions. For example, for a function with a large, monotonic drop between the starting point and the minimum, it would require a high noise level to cause the algorithms to fail. If the terrain between the

starting point and the minimum is more complex, a small noise level could suffice to defeat the algorithm.

7.3.2 RCDS algorithm

Powell's method is powerful for the optimization of smooth functions. However, its performance is sensitive to noise, as seen in Figure 7.7. A close examination of the algorithm reveals that the negative impact of noise comes in through the 1-dimensional optimizer. For both the golden section method and the inverse quadratic interpolation method, the first step is to bracket the minimum within a finite zone. The bracketing procedure relies on the comparison of function values at the boundary points and a point inside. Noise in the function values can change the comparison results and lead to a false bracket. Noise can also change the next step of the 1-dimensional optimizer. For the golden section method, it changes the selection of the subdivision to move in; for the inverse interpolation method, it changes both the prediction of the trial solution and the selection of the subdivision.

The 1-dimensional optimizers suffer from noise because they have no consideration of the existence of the noise. Small changes to the optimizers could significantly suppress the noise effect. The robust conjugate direction search (RCDS) [57] algorithm makes the 1-dimensional optimizer aware of the noise level and takes measures to ensure the decisions in the bracketing and interpolation steps are valid under noise. The resulting 1-dimensional optimizer is called the robust line optimizer. It is substantially more robust against noise.

Starting from the present solution, $g_0 = g(\alpha = 0)$, the robust line optimizer first finds the boundary in one direction (say, the positive direction) by sampling new solutions with increasingly longer steps. The step length is increased by a factor of 1.618 after each new data point, but is capped at a certain level, say, 0.1 (note normalized coordinates are used). During the search it keeps updating the current minimum, g_{min}. The search continues until the boundary of the parameter range is reached, or a point $\alpha = b$ is found which satisfies

$$g(b) > g_{min} + \kappa\sigma, \tag{7.32}$$

where σ is the noise sigma and κ represents the required confidence level, which can usually be chosen to be 3. After the bracket boundary at the positive direction is determined, it then starts from $\alpha = \alpha_{min}$ and goes in the opposite direction to find the lower bracket end, a.

After the minimum is bracketed, all the sample points on the line are fitted to a parabola. If very large steps are taken and hence there are large gaps between the sample points, intermediate points inside the gaps can be evaluated. The fitted parabola is used to predict the position of the minimum, which is then sampled as the next trial solution. Figure 7.9 illustrates the procedure executed by the robust 1-dimensional optimizer, in which the indices beside the sample points indicate the order in which the date points are evaluated.

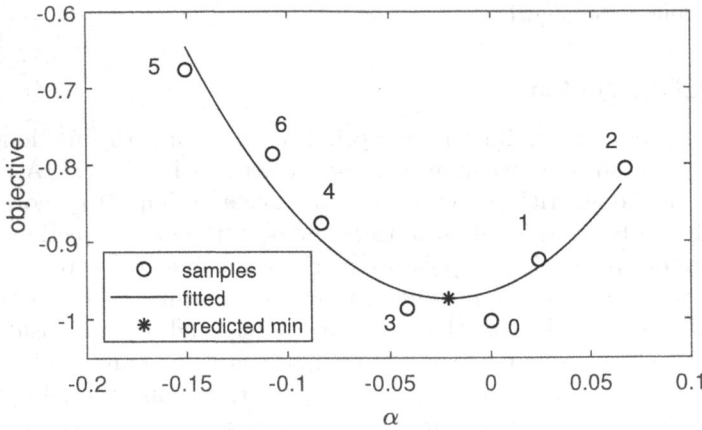

Figure 7.9 Illustration of the robust line optimizer. The sample point indices represent the order of data taking, starting with the initial solution (point 0).

In this case, 7 data points are taken, with point 0 for the initial solution, points 2 and 5 define the bracket boundary, and point 6 is an extra sample point to fill the gap.

The RCDS algorithm combines the management of the conjugate direction set of Powell's method and the robust line optimizer described in the above. Because of its awareness of noise and the measures taken to avoid being misled by the noise, the performance of the algorithm is not sensitive to noise. Figure 7.10 shows the test results of RCDS with the analytic function in Eq. (7.31). For all three noise levels, the algorithm can successfully find the minimum. The left plot shows that the minimum function value typically reaches the level $f(\mathbf{x}) < 0.01$ within 60 evaluations, much faster than the other algorithms shown previously.

7.3.3 RSimplex algorithm

An effort has also been made to modify the Nelder-Mead simplex method in order to improve its performance under noise, which resulted in the robust simplex algorithm [55]. The modifications are based on the observations on how noise impacts the operations of the Nelder-Mead simplex method.

An important step in an iteration of the Nelder-Mead simplex algorithm is to find the vertex with the worst function value – the subsequent operations are a search along the line from this vertex to the center point of its opposite simplex face. The worst vertex is chosen by sorting the function values on all vertices. The sorting results can be changed by noise in the function values. The decision of accepting or rejecting the result of an operation (i.e., reflection,

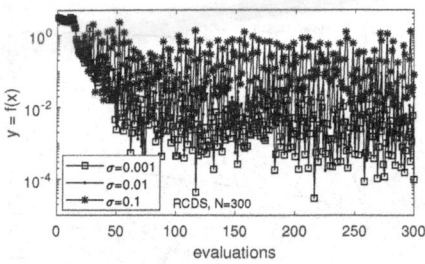
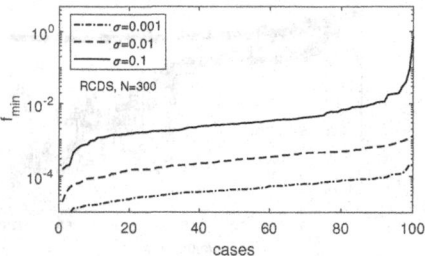

Figure 7.10 Testing of the RCDS method with function $f_1(\mathbf{x})$ defined in Eq. (7.31) and three levels of noise. The algorithm can find the minimum reliably and efficiently. Left: the cases whose final minimum rank the 30th among 100 runs; right: the sorted final minima of the 100 runs.

expansion, and contraction) also depends on the comparison of function values. Therefore, improving the reliability of function value comparisons would have a big impact.

With Gaussian noise, the function value sampled at any point, $y = \mu + \xi$, follows a normal distribution, $\mathcal{N}(\mu, \sigma^2)$, where $\mu = f(\mathbf{x})$ and σ is the standard deviation of the noise variable ξ. The comparison of function values at two points is to determine the sign of $\mu_1 - \mu_2$ using $y_1 - y_2$, where subscripts stand for the two points. The comparison result would be usually correct if $|\mu_1 - \mu_2| \gg \sigma$. Conversely, it would be often incorrect if $|\mu_1 - \mu_2|$ is smaller or comparable to σ. To ensure reliable comparisons, it is desirable to maintain an appropriate simplex size. It will also help if multiple samples are taken at each point and the average values, $\bar{y}_1$ and $\bar{y}_2$, are used for comparison. If the numbers of samples at the two points are N_1 and N_2, respectively, the distribution of $\bar{y}_1 - \bar{y}_2$ is

$$\mathcal{N}(\mu_1 - \mu_2, \Sigma^2), \text{ with } \Sigma^2 = (\frac{1}{N_1} + \frac{1}{N_2})\sigma^2.$$

The distribution can be used to determine the number of samples required in order to obtain reliable comparison results, based on the observation that $\bar{y}_1 - \bar{y}_2$ and $\mu_1 - \mu_2$ have the same sign with the probability of $\frac{1}{2} + \frac{1}{2}\text{erf}(\frac{|\mu_1 - \mu_2|}{\sqrt{2}\Sigma})$, where $\text{erf}(\cdot)$ is the Gaussian error function. If we use $|\bar{y}_1 - \bar{y}_2|$ as an estimate of $|\mu_1 - \mu_2|$, we can require

$$|\bar{y}_1 - \bar{y}_2| > M_1 \sigma \sqrt{\frac{1}{N_1} + \frac{1}{N_2}}, \tag{7.33}$$

for a certain level of confidence in the comparison results. For example, for $M_1 = 1.4$, the comparison results would be correct with a 92% probability. In the RSimplex method, the number of samples for each evaluated solution is recorded. More samples will be taken if necessary in order to resolve the

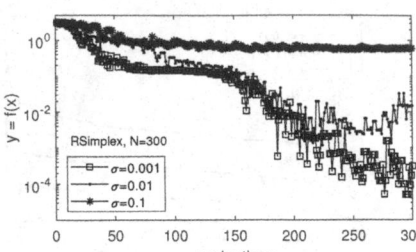 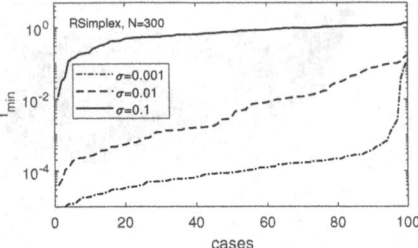

Figure 7.11 Testing of the RSimplex method with function $f_1(\mathbf{x})$ (Eq. (7.31)) and three noise levels. Left: typical cases (ranking the 30th) for the three noise levels; right: sorted final minimum values for 100 runs.

ambiguity of a comparison result. However, an upper limit is given to the number of samples to avoid excessive refinement.

A number of other modifications are also made to the original simplex method. When there are several contenders for the worst vertex, the algorithm will select one from them and perform the usual sequence of operations; if it does not lead to the replacement of the vertex and the termination of the present iteration, it will sequentially use the other candidates as the worst vertex.

If the sequence of operations along the line does not lead to the replacement of the worst vertex, the solution at the center of its opposite simplex face is also evaluated and the five points on the line are fitted with a quadratic function. The fitting result is used to select between the inside contraction and the outside contraction point.

The shrink operation leads to a substantial decrease of the simplex size and can cause the simplex method to prematurely converge to a non-optimum. This is especially the case for noisy functions. To prevent the shrinking of the simplex size to a level when the noise dominates the comparison of function values, the RSimplex method allows the shrink operation only if the difference between the maximum and the minimum vertices is larger than $M_2\sigma$, where M_2 is a constant which can be set to 2 or larger. The algorithm can optionally rebuild the simplex around the minimum vertex after the other operations fail to reduce the minimum or the maximum vertex values.

Figure 7.11 shows the test results for the RSimplex method with the test function $f_1(\mathbf{x})$. A substantial improvement is made by the RSimplex from the N-M simplex method for most cases under medium or high noise levels. However, at low noise level, the original simplex and RSimplex may have similar performance in terms of the final minimum achieved. The original simplex method could converge faster sometimes as RSimplex may take extra evaluations to confirm certain comparisons.

7.3.4 Performance comparison for single-objective algorithms

The tests with function $f_1(\mathbf{x})$ in Eq. (7.31) demonstrate the impact of noise on the performances of a few selected single-objective optimization algorithms. Figure 7.12 summarizes the test results. It shows the comparison of the sorted minimum function values found in 300 evaluations for 100 runs by the algorithms under three noise levels. Under low noise ($\sigma = 0.001$), almost all algorithms can locate the minimum (with final minimum $f_{\min} < 0.1$), except the N-M simplex method occasionally fails. Under medium noise ($\sigma = 0.01$), the gradient descent method and the N-M simplex often fail to achieve $f_{\min} < 0.1$, while the others usually can. When the noise level is high ($\sigma = 0.1$), all other algorithms fail to reach $f_{\min} < 0.1$, except for the RCDS method. The RCDS is the least sensitive to noise among all tested algorithms. The RSimplex method outperforms the original simplex in all three noise levels, although it falls behind the RCDS method.

To further characterize the performances of the algorithms, additional tests are done with the Rosenbrock function [100]. The Rosenbrock function is a non-convex function, given in the form

$$f_2(\mathbf{x}) = \sum_{i=1}^{N-1} \left[100(x_{i+1} - x_i^2)^2 + (1 - x_i)^2 \right], \tag{7.34}$$

where N is the number of dimensions in $\mathbf{x}$. In the test we choose $N = 4$, in which case the Rosenbrock function has one global minimum located at $\mathbf{x}_m = (1, 1, 1, 1)^T$, with $f_2(\mathbf{x}_m) = 0$, and a local minimum at $(-1, 1, 1, 1)^T$ with the function value of 4. The parameter ranges are chosen to be the same as the tests for the function in Eq. (7.31), with $x_i \in [-2, 2]$ for $i = 1$-4. The initial solution is also chosen to be the same, with $\mathbf{x}_0 = (-0.5, 0, -0.5, 0)^T$, where the function value is $f(\mathbf{x}_0) = 43$.

The tests are run with three noise levels, with $\sigma = 0.001$, 0.01, and 0.1. The number of evaluations is limited to 500 for all algorithms except for the GP optimizer. For the GP the number of evaluations is limited to 300 as it becomes too time consuming and there is no clear gain with more evaluations (see Figure 7.14). The final minimum function values achieved in 100 runs are sorted and plotted in Figure 7.13 for all algorithms and the three noise levels. With low noise ($\sigma = 0.001$), the RSimplex and N-M simplex methods have similar performance, and both of them outperform the other methods. With medium noise ($\sigma = 0.01$), the RSimplex shows better resistance to noise than the N-M simplex method. The RCDS method also performs better than the original simplex method for most of the cases. At a high noise level ($\sigma = 0.1$), while RSimplex is still better than the original simplex method, it is not as good as the RCDS method. The performance of the RCDS method for the three noise levels was similar, but its efficiency was not high. This is due to the nature of the Rosenbrock function, for which the valley in the parameter space leading to the minimum is banana-shaped, such that the direction toward the minimum is constantly changing. Figure 7.14 shows the convergence histories

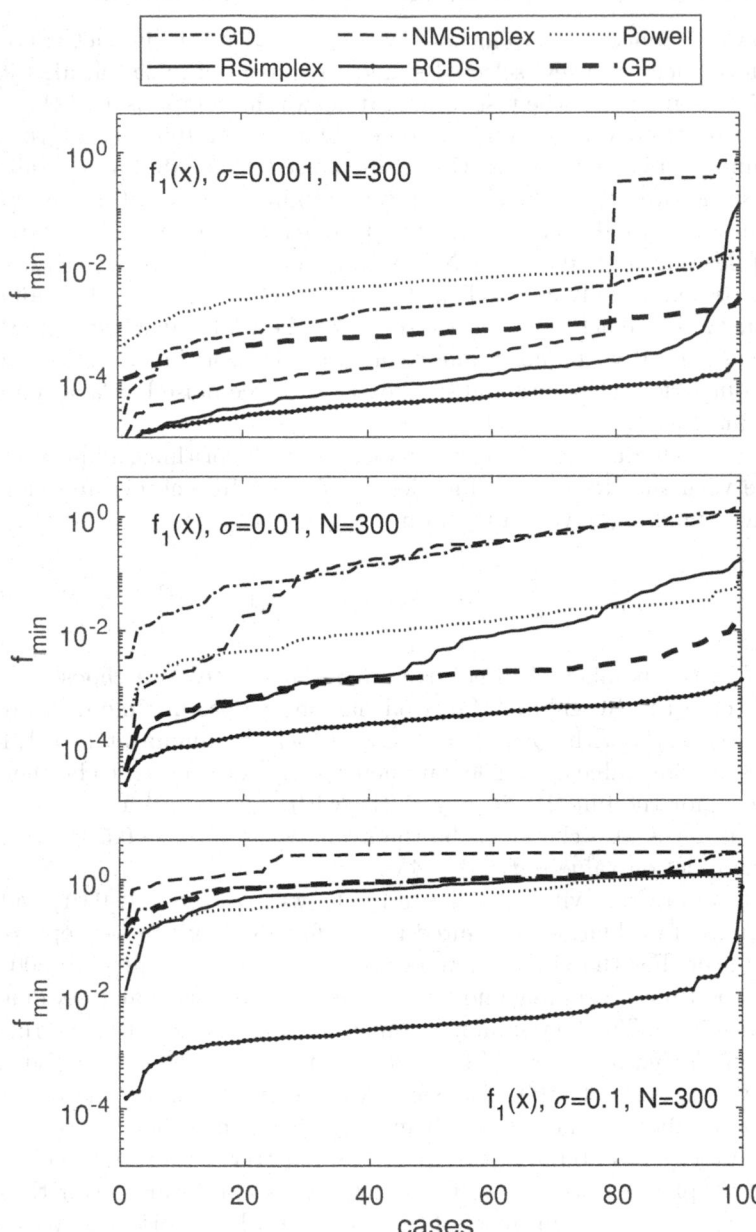

Figure 7.12 Comparison of optimization performances for selected algorithms using the function in Eq. (7.31). The noise levels are $\sigma = 0.001$, 0.01, and 0.1 from top to bottom. The sorted final minima for 100 runs are shown for each algorithm.

of the RCDS, RSimplex, N-M simplex, and GP methods for the Rosenbrock-4 function, using the cases ranking the 30th out of the 100 runs for each noise level as examples.

While the Rosenbrock function is challenging for the RCDS method in terms of the convergence efficiency, real-life online optimization applications typically do not have such a behavior in the parameter space. For many problems, the design operation condition corresponds to an extremum of the objective function, around which the performance behaves as a quadratic function of the control parameters. The RCDS method would be an ideal algorithm to bring the machine toward the design performance in such cases, especially when the starting point is close to the optimum.

7.3.5   Testing of multi-objective optimization with stochastic algorithms

Deterministic optimization algorithms usually converge to the nearby local minimum. If in an application the objective function has local minima that could intercept the convergence path of the deterministic algorithms toward the global minimum, stochastic algorithms can be used to look for the global minimum. Stochastic algorithms may not be as efficient, but typically have better ability to overcome the attraction of local minima.

The NSGA-II genetic algorithm and the particle swarm optimization (PSO) algorithm are two popular stochastic algorithms that are used for accelerator designs. The MG-GPO algorithm is a new stochastic optimization method. These algorithms naturally apply to multi-objective problems. We tested their performances under noise with a multi-objective problem, using the two objective functions, $f_1(\mathbf{x})$ in Eq. (7.31) and the Rosenbrock-4 function. The parameter range is the same as used in the previous tests. For all three methods, the population of solutions is set to $N = 50$ and the algorithms are run for 40 generations in the tests. The initial population is randomly chosen from the entire parameter space with a uniform distribution.

In the NSGA-II setup, 90% of the new solutions are generated through crossover and the 10% by mutation. The control parameters for random number generation in Eq. (7.19) and Eq. (7.21) are set to $\mu_c = \mu_m = 20$.

For the PSO setup, the weight factors for velocity calculation in Eq. (7.23) are set to $w = 0.4$ and $c_1 = c_2 = 1$. The components of all initial velocities are drawn from the uniform distribution between 0 and 0.1.

The MG-GPO setup assumes $\theta = 0.4$ for the correlation length parameter. The GP-LCB acquisition function uses $\kappa = 0.5$. There are $40N$ trial solutions, half generated with crossover and the other half with mutation, before the selection by the GP model is applied.

Three levels of noise are applied to the function evaluations as was done in the single-objective tests. About 2000 solutions are evaluated for each algorithm, out of which the best 100 solutions are selected through non-dominated sorting. The results are shown in Figure 7.15 for NSGA-II and PSO. The final fronts of MG-GPO (not shown) are similar to that of PSO. At the low noise

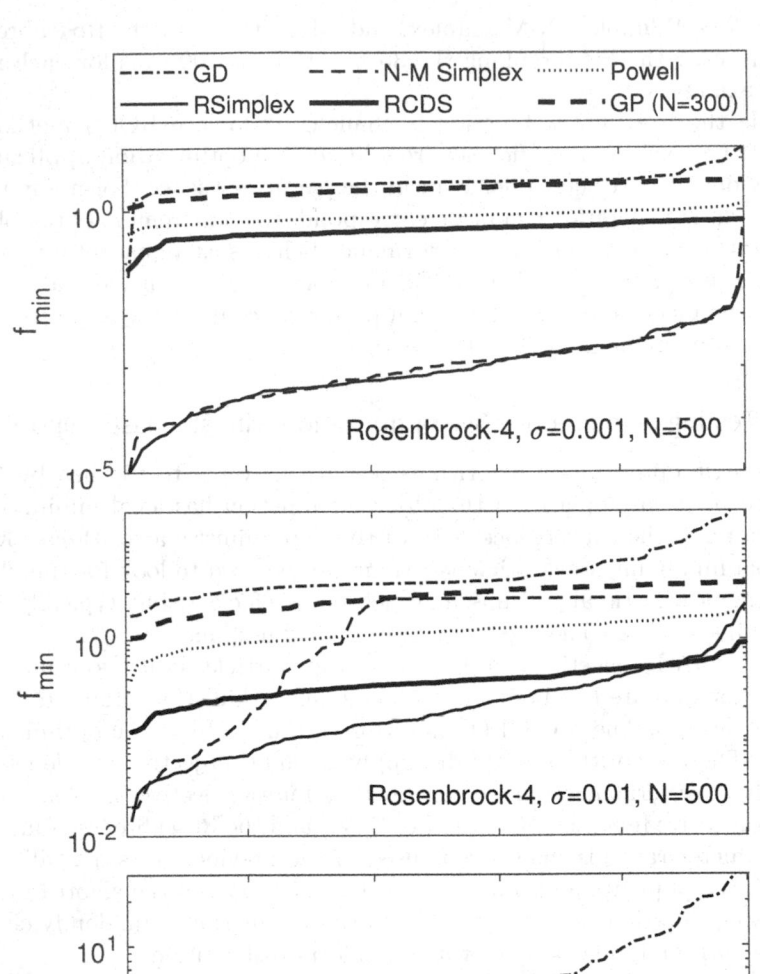

Figure 7.13 Comparison of optimization performances for selected algorithms using the Rosenbrock function with 4 variables. The sorted final minima of 100 runs are shown for the noise levels $\sigma = 0.001, 0.01$, and 0.1 (from top to bottom).

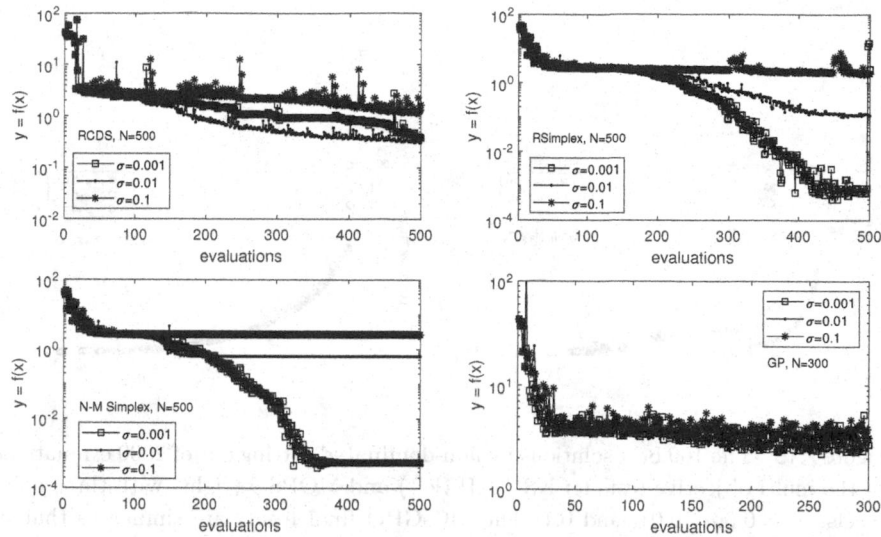

Figure 7.14 The convergence histories over 500 function evaluations of RCDS (top left), RSimplex (top right), and N-M simplex (bottom left) and over 300 evaluations for GP (bottom right) for the Rosenbrock-4 function with three noise levels.

level ($\sigma = 0.001$), all algorithms converge toward the Pareto front, although the NSGA-II front is narrow and incomplete. With higher noise ($\sigma = 0.01$ or 0.1), the NSGA-II method fails to converge to the Pareto front, while the PSO and MG-GPO algorithms converge to the same front for the medium or high noise levels.

The Figure 7.16 left plot shows the minimum values for the two objective functions over the course of optimization for the three algorithms for the low-noise case ($\sigma = 0.001$). The convergence speed of MG-GPO is much faster than PSO, which is in turn much faster than NSGA-II. The cause of the slow convergence for NSGA-II is the low diversity of its new trial solutions. The diversity can be measured by the distribution of the crowding distance, here defined as the distance of a solution to its nearest neighbor in the parameter space among all previous solutions. The right plot shows the distribution of the crowding distances for the three algorithms. Interestingly, the diversity of MG-GPO is even lower than NSGA-II. It still leads to high efficiency because the MG-GPO solutions are selected with the posterior model and are not entirely random.

Since MG-GPO and PSO are not sensitive to noise and are more efficient than genetic algorithms, they are preferred for problems where stochastic algorithms are required to search the global optima.

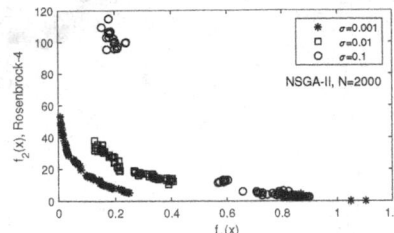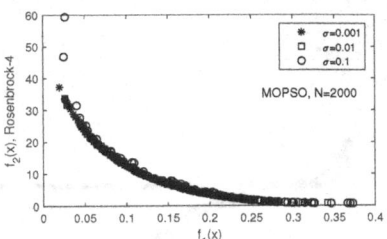

Figure 7.15 The 100 best solutions by non-dominated sorting out of 2000 evaluations in the multi-objective tests for NSGA-II (left) and MOPSO (right) with three noise levels, $\sigma = 0.001$, 0.01, and 0.1. The MG-GPO final fronts are similar to that of PSO. Noise is not included in the function values shown in the plots.

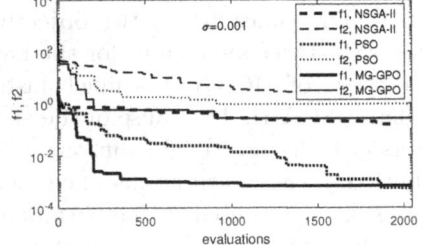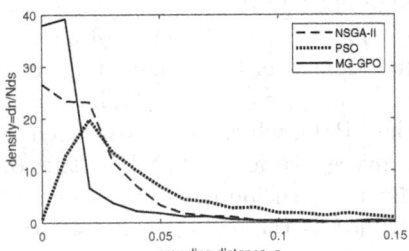

Figure 7.16 Left plot: the minimum values for $f_1(\mathbf{x})$ (see Eq. (7.31)) and $f_2(\mathbf{x})$ (Rosenbrock-4) over the course of the optimization run for NSGA-II, PSO, and MG-GPO (left) for $\sigma = 0.001$. Right plot: the corresponding distribution of the crowding distance of all evaluated solutions, $\frac{dn}{Nds}$, where N is the total number of solutions.

## CHAPTER 8

Application of beam-based optimization

CONTENTS

In the previous chapter we discussed the general considerations of online optimization and the various online optimization algorithms. In this chapter we will demonstrate the application of online optimization to real-life accelerator problems. These experiments may serve as examples for setting up online optimization for other problems.

The applications to be described are

- Linac-to-booster (LTB) trajectory steering,

- Storage ring injection kicker bump matching [57],

- Storage ring coupling minimization [57], and

- Storage ring dynamic aperture optimization [60].

All experiments were conducted on the SPEAR3 accelerator complex. The RCDS algorithm was the main method used for these experiments.

There have been many applications of online optimization on other accelerators, including experiments similar to the ones to be discussed here. Some of these experiments will also be described.

8.1 LINAC-TO-BOOSTER TRAJECTORY STEERING

The linac-to-booster (LTB) transport line is the part of the SPEAR injector that connects the 120-MeV linac to the Booster synchrotron. The steering of the beam trajectory in the LTB is important for achieving a good capture efficiency for the beam injected into the Booster. It is frequently tuned during operations.

The Booster has a finite transverse acceptance. For the best capture efficiency, the LTB beam trajectory at the injection point should have the proper position and angle such that beam is at the center of acceptance in both transverse planes. There are a number of steering knobs that can affect the trajectory at the end of the LTB, including corrector magnets in the linac, corrector magnets and the trim coils on the bending magnets in the LTB, the injection kicker, the injection septum, and the power of the linac klystron K3. The K3 power knob changes the beam energy and in turn the trajectory due to dispersion in the transport line.

A pair of steering magnets located near the end of the transport line are chosen for the steering experiment for each plane. In the horizontal plane, the trim coils on the B3 bending magnet (B3trim) and the injection septum are used. The correctors COR3V and COR5V are used for the vertical plane. The two knobs in each plane can effectively change both the position and angle coordinates of the beam at the injection point.

The tuning of one steering magnet shifts the beam trajectory along a line in the (x, x') or (y, y') phase space. Ideally, the lines traced by the two knobs in each plane should be orthogonal; in such a case, the steering of one knob is independent of the other. Such knobs can be formed by combining the two steering magnets with a proper ratio of the strengths. For example, one combined knob is for the x coordinate, and the other for the x' coordinates, and similarly for the vertical plane. However, in this experiment we simply used the setpoints of the individual magnets as knobs. The parameter ranges are $(-1, 1)$ A for the three steering magnets, B3trim, COR3V, and COR5V, and $(-0.5, 0.5)$ A for the injection septum, relative to the initial setpoints. The ranges are large enough such that the deviation of any knob can significantly reduce the capture efficiency before reaching the limit.

The Booster beam current (the monitor is referred to as the Q-meter) measured near the end of the ramping cycle is used as the optimization objective. The intensity of the linac beam is stable in a short period of time. The injection loss only occurs near the beginning of the cycle. Therefore, any change to the capture efficiency will be reflected on the Q-meter reading.

After the knobs are changed, the code waits for 3 seconds for the magnets to settle to the new setpoints. The beam is then turned on and monitored for 2 seconds. Since the injector runs with a 10-Hz repetition rate, there are 20 valid Q-meter data points. The average value, with a minus sign added, is used as the objective function. The minus sign is inserted to make maximizing the beam current a minimization problem.

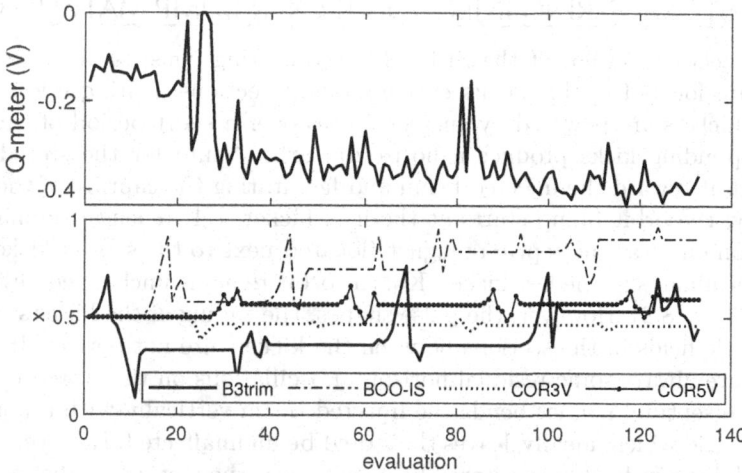

Figure 8.1 Optimization of the beam trajectory at the end of the LTB transport line to increase the capture efficiency of the Booster, using two horizontal steering magnets and two vertical steering magnets. Top plot: Q-meter reading as a measure of the Booster beam intensity at a beam energy near 3 GeV; bottom plot: the normalized values of the four tuning knobs. The RCDS algorithm was used.

At the beginning of the experiment, the noise sigma of the objective function was evaluated by taking 20 measurements, while the LTB was in the standard operation setting. The average value for the Q-meter was -0.354 V (beam intensity in uncalibrated, raw data) and the noise sigma was $\sigma = 0.009$ V.

In the experiment, two corrector magnets upstream of the tuning knobs in use were first changed to reduce the Q-meter reading to about -0.15 V. The RCDS algorithm was used to tune the two pairs of steering magnets. The results are shown in Figure 8.1. After the first iteration (about 30 evaluations), the Q-meter was restored to the level of -0.35 V. In two more iterations, the performance continued to slowly increase. The Q-meter reached -0.426 V for the best solution.

The trajectory with the best capture performance can be recorded with the BPMs and is used as the target for a trajectory correction program, which can maintain the LTB trajectory. However, the ideal trajectory may drift with time as the Booster beam orbit at the injection point may change. Since one iteration of RCDS run for the LTB steering takes about 2.5 minutes, and the injection for SPEAR3 occurs every 5 minutes, it is possible to tune the trajectory for one iteration between two fills.

8.2 STORAGE RING INJECTION KICKER BUMP MATCHING

The injection system of the SPEAR3 storage ring consists of three kickers that are located in three consecutive straight sections. During injection, the three kickers are powered by current pulses over a short period of time. The corresponding kicks produce a horizontal orbit bump for the stored beam, moving it toward the injected beam and facilitating the capture of the latter. Ideally, the orbit bump starts at the first kicker, K1, reaches the maximum position offset at the septum magnet (located next to the second kicker, K2), and terminates at the last kicker, K3; the orbit remains unchanged beyond K3 (see Figure 8.2). However, the pulse shapes, the timing of the kickers, and the magnetic fields in the sections between the kickers are not perfect. In reality, there are always some residual horizontal oscillations on the stored beam.

The septum magnet bends the injected beam vertically with a horizontal dipole field which, ideally, leaves the stored beam unaffected. However, there is a small leakage field in the stored beam side near the metal wall that separates the two beams. When kicked to the orbit bump, the stored beam sees the leakage field and gets a vertical kick, causing vertical betatron oscillations.

The horizontal and vertical residual oscillations persist in a duration of about two damping times (~10 ms). The oscillations may become a visible perturbation to the user experiments if the amplitudes are large. Therefore, it is desirable to minimize the residual oscillations due to the kickers and the septum leakage field. The horizontal oscillation can be minimized by matching the amplitudes, pulse widths, and the timing of the three kicker pulses. The vertical oscillation can be minimized by compensating the leakage field with the two skew quadrupoles located within the orbit bump. The desired kicker amplitude at the target beam bunch is determined by the parameters of the K1 pulse. Hence the K1 parameters are not to be varied. Instead, the parameters

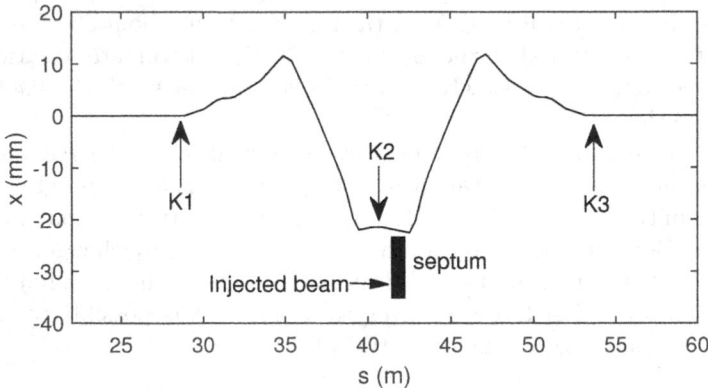

Figure 8.2 Illustration of the injection kicker bump for SPEAR3. The locations of the three kickers and the septum manget are indicated.

of K2 and K3 are varied to match the K1 parameters. There are 8 control parameters in total: the pulse voltage (amplitude), the pulse width, and the pulse timing of K2 and K3, and the two skew quadrupoles [57].

Betatron oscillations can be measured with a turn-by-turn BPM. Because of the different pulse shapes of the three kickers and the potential time shifts between the kickers, for a given kicker setting, the amplitudes and phases of residual oscillations at different bunches in a bunch train are not the same. Since the turn-by-turn BPM measures the average position of all bunches, it could happen that some bunches are oscillating in opposite phases and their motion is not detected by the BPM. A likely case is for the timing of one kicker to be shifted from the perfectly matched setting, which causes the bunches in the head and the tail of the kicker pulse to see net kicks of opposite signs. The goal of kicker bump matching is to minimize the average of oscillations at all bunches, as measured by the sum of squares of their amplitudes. To avoid the cancellation in the measured beam motion, a bunch train much shorter than the pulse width can be used. To measure the residual oscillations at a different position in the bunch train, the kicker timing is shifted to align with the certain bucket. The objective of optimization is the sum of the oscillation amplitudes of the short bunch train with a few kicker timing selections. For the 372 buckets in SPEAR3, a bunch train of 186 bunches can be used and the kicker pulses can be aligned with buckets 1, 93, 187, and 280.

The objective function is defined as

$$f = \frac{1}{4} \sum_{i}^{4} (\sigma_{ix} + 3\sigma_{iy}), \tag{8.1}$$

where subscript $i = 1$–4 indicates the four timing selections, $\sigma_{x,y}$ are the rms orbit of the first 256 turns of the horizontal and vertical oscillations, respectively. The vertical oscillation is given a high weight because user experiments are more sensitive to vertical orbit disturbance, due to the small vertical emittance of the beam.

The parameter range was [0.5, 1.9] kV for the kicker amplitudes, [0, 100] ns for kicker pulse delays, [600, 800] ns for kicker pulse widths, and [−15, 15] A for the skew quadrupoles. The noise sigma was first measured by taking 20 readings under the initial setting and was found to be $\sigma_f = 4.3$ μm.

In the experiment the eight parameters were first intentionally shifted to create a poorly matched kicker bump. The RCDS algorithm was used to minimize the objective function. All initial search directions are simply the parameter axes. It took three iterations (each iteration covers 8 directions) in about 180 total function evaluations. The history of the optimization run, including the evolution of both the objective function and the control parameters, is shown in Figure 8.3. The objective was reduced from 1300 μm to 300 μm in about 80 evaluations. In Figure 8.4 the left plot shows the measured turn-by-turn betatron oscillations before and after the optimization. After optimization, the horizontal and vertical rms orbits in the first 256 turns are about

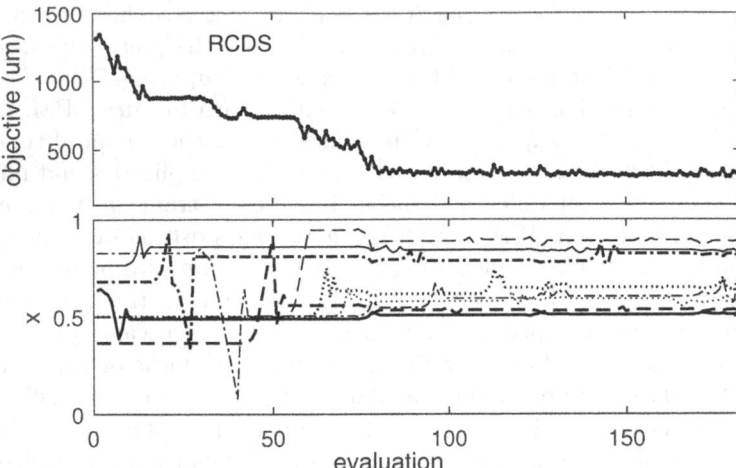

Figure 8.3 Minimization of the residual oscillations of the stored beam due to the injection kicker bump with the RCDS method: evolution of the objective function (top) and the eight normalized parameters (bottom) are shown.

150 μm and 50 μm, respectively. The right plot shows the kicker pulses for the optimized solution.

The Nelder-Mead simplex method was also used to minimize the residual oscillations in the same experiment. The evolution of the objective function and the knob parameters are shown in Figure 8.5. While the algorithm initially reduced the objective function efficiently, it converged to a non-optimum prematurely. The objective function did not reach the same value as the RCDS method. The size of the initial simplex is 5% of the parameter range. In such cases, using a larger initial simplex size, or relaunching the simplex from the new solution could help.

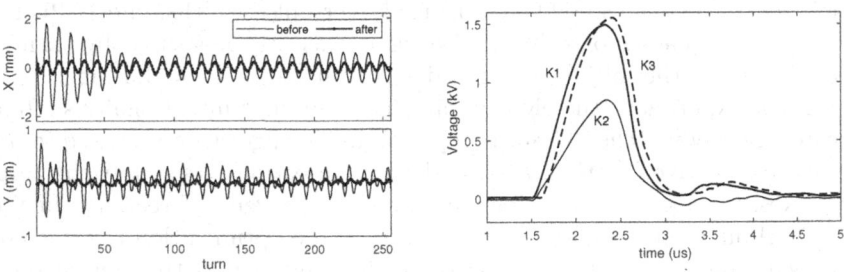

Figure 8.4 Left: the residual oscillations before and after the RCDS optimization; Right: the pulses for the three kickers (note the pulse delays may include errors due to signal cable lengths).

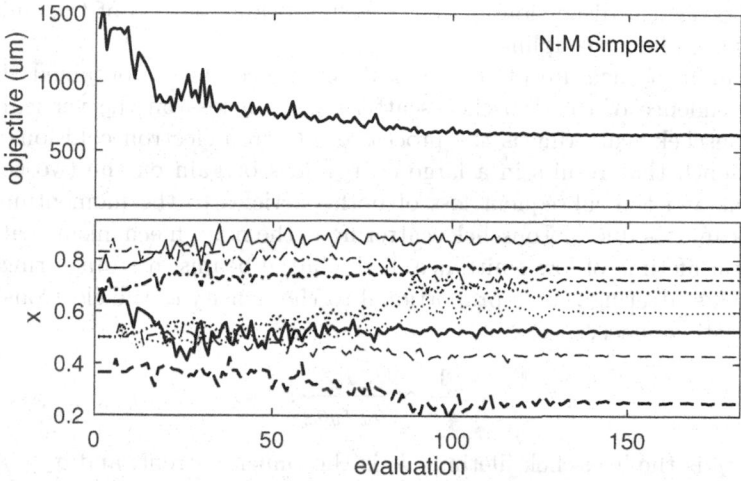

Figure 8.5 Minimization of the residual oscillations with the N-M simplex method. Top: the objective function; bottom: the eight normalized parameters.

8.3 VERTICAL EMITTANCE MINIMIZATION

An ideal electron storage ring would have a very small vertical emittance, arising only from the excitation by the vertical photon divergence within the angular range of $\theta \lesssim \frac{1}{\gamma}$. In reality, there are lots of error sources that cause linear coupling between the horizontal and vertical planes and the spurious vertical dispersion, both of which contribute to the vertical emittance. The vertical to horizontal emittance ratio can easily reach a few percent or higher if no action is taken to compensate the errors.

Both the linear coupling and the vertical dispersion can be corrected with skew quadrupoles. In Chapter 6 we discussed coupling correction with various beam-based correction techniques. The vertical emittance can also be minimized with beam-based optimization. The latter approach is simple to implement: one only needs to provide a measure of the vertical emittance to serve as the objective function; the same skew quadrupoles used in coupling correction are to be used as the optimization knobs.

The vertical emittance can be determined through vertical beam size measurements, e.g., using a pinhole camera or an interferometer to measure the radiated photon beam at a beamline. As the projected emittance may vary with locations due to linear coupling, minimization of the vertical beam size at one location could potentially result in a change of the local coupling angle rather than a global reduction of the vertical emittance. It is desirable to have simultaneous beam size measurements at multiple locations, which is, however, usually not available. Nonetheless, minimization of the vertical beam size at one location is still useful since it can target the emittance contribution from the vertical dispersion as well as a significant fraction of the emittance coming from linear coupling.

An indirect measure of the vertical emittance can be obtained through the dependence of the Touschek scattering beam loss on the vertical beam size. Touschek scattering is the process of electron-electron collisions in the beam bunch that results in a large energy loss or gain on the two colliding electrons and the subsequent loss of both particles to the momentum aperture. Beam loss due to Touschek scattering is the main mechanism that limits the beam lifetime of low emittance, high charge beams in storage rings. The Touschek scattering rate is proportional to the density of the electrons in the configuration space,

$$\frac{1}{\tau_T} \propto \frac{I_b}{\sigma_x \sigma_y \sigma_z}, \tag{8.2}$$

where τ_T is the Touschek lifetime, I_b is the bunch current, and σ_{xyz} are the beam sizes in the three dimensions. Since σ_x and σ_z are mostly not affected by skew quadrupole variations, any change of the Touschek loss rate by skew quadrupoles must come through a change of the average inverse vertical beam size, $\langle \frac{1}{\sigma_y} \rangle$, throughout the ring.

The Touschek loss rate can be measured by observing the beam current change, ΔI, over a short period of time, Δt, for a beam with a high bunch charge. For a Touschek loss dominated beam, we have $\Delta I = -\frac{I \Delta t}{\tau_T}$. Hence, the objective function may be defined as

$$f(\mathbf{x}) = \frac{I_0^2 \Delta I}{I^2 \Delta t} \propto -\langle \frac{1}{\sigma_y} \rangle, \tag{8.3}$$

where I_0 is a reference current. Note the minimization of the objective function is equivalent to minimizing the vertical emittance.

Vertical emittance minimization with beam-based optimization has been applied to the SPEAR3 storage ring in both simulation and experiments [57, 115]. This problem is ideal for testing online optimization algorithms as it is a real-life application with sufficient challenges in the number of knobs, the noise level, and the parameter space complexity. In the following both simulation and experimental tests are discussed.

8.3.1 Simulation

With a storage ring lattice model, the equilibrium distribution of the electron beam can be obtained by calculating the one-turn transfer matrix with radiation damping included and the radiation-induced diffusion matrix integrated throughout the ring and solving for a self-consistent second order moment matrix [90]. The horizontal and vertical emittances can be readily calculated from the second order moment matrix. This calculation has been implemented in the lattice modeling code Accelerator Toolbox (AT) [114], which is used in this simulation.

The SPEAR3 storage ring has 72 sextupole magnets, each of them has a set of skew quadrupole coils. Only 15 of these skew quadrupoles are powered. Not using the two skew quadrupoles within the injection kicker bump (see Section 8.2), there are 13 skew quadrupoles available for vertical emittance minimization. In simulation, skew quadrupole errors are added to 29 sextupoles, not including the 13 knobs. The coupling ratio (i.e., $r = \epsilon_y/\epsilon_x$) is 0.88% initially. Assuming that for a 500 mA total beam current, the gas scattering lifetime is 40 hrs and the Touschek lifetime is 10 hrs for the coupling ratio of 0.2%, the total lifetime can be calculated with

$$\frac{1}{\tau}[\text{hr}] = \frac{1}{40} + \frac{I}{10I_0}\sqrt{\frac{0.002}{r}}. \tag{8.4}$$

The current loss over Δt is calculated with $\Delta I = -I\frac{\Delta t}{\tau} + \sqrt{2}\sigma_I\xi$, where σ_I is the error sigma of beam current measurement, ξ is a random number drawn from the Gaussian distribution, $\mathcal{N}(0, 1)$. The random term is added to simulate the measurement error. The objective function is calculated with Eq. (8.3), using $I_0 = 500$ mA and scaled to the beam loss over one minute. The loss rate for the initial lattice is -0.61 mA/min. Assuming $\sigma_I = 0.002$ μA, and for the beam current change measured over a period of $\Delta t = 6$ seconds, the noise sigma for the objective function is 0.03 mA/min.

For the application of the RCDS method, it is preferable to provide a conjugate direction set as the initial directions. Although the method has the ability to build up a conjugate direction set from the convergence history, it takes many iterations to replace all the directions. In online applications, there is typically not enough time to wait for the conjugate direction set to emerge. The initial conjugate direction set may be calculated with a model. The final direction sets of past RCDS runs may also be used. For the vertical emittance minimization problem, we use the Jacobian matrix of the orbit response matrix fitting with respect to the skew quadrupoles to calculate the conjugate directions. Each column of the Jacobian matrix consists of the derivatives of all orbit response matrix elements (including dispersion functions in both planes) with respect to one skew quadrupole. The off-diagonal matrix elements and the vertical dispersion dominate the response of skew quadrupoles. Because the Jacobian matrix contains sufficient information of the dependence of the distributions of the linear coupling and the vertical dispersion on the skew quadrupoles, it can represent the functional dependence of the vertical emittance on the skew quadrupoles. By the use of singular value decomposition (SVD) of the Jacobian matrix, $\mathbf{J} = \mathbf{USV}^T$, the conjugate direction set, given by the column vectors in matrix $\mathbf{V}$, can be found. The combined knobs represented by the directions in $\mathbf{V}$ are ordered by the sensitivity in changing the linear coupling and the vertical dispersion, and in turn the vertical emittance. The left plot in Figure 8.6 shows the singular values of the Jacobian matrix and the square root of the coupling ratio caused by a fixed step in each knob. The right plot shows the first two conjugate directions.

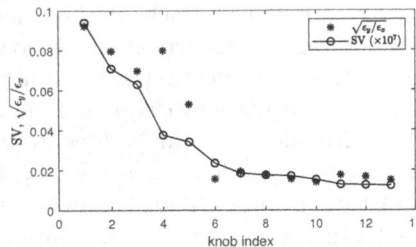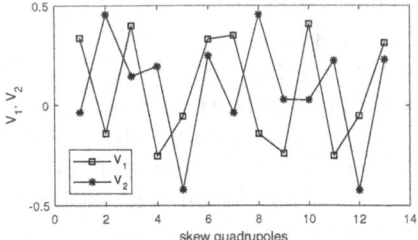

Figure 8.6 Left: the SVs of the Jacobian matrix of the orbit response matrix with respect to the 13 skew quadrupoles and the sensitivity of the vertical emittance to the conjugate directions (in term of $\sqrt{\frac{\epsilon_y}{\epsilon_x}}$ for a fixed step) for SPEAR3; right: the first two conjugate directions, v_1 and v_2.

The range of the skew quadrupole gradient is set to $K_1 \in [-0.3,\ 0.3]$ m^{-2}. The magnet lengths may be 21 cm or 25 cm. The initial solution is with all 13 knobs turned off. The RCDS method is applied to the problem with the calculated conjugate direction set or using the unit vectors as the initial direction set. The Nelder-Mead simplex and the RSimplex methods are also tested. The initial simplex side length is 5% of the full range for each knob. The algorithms run 100 times each with different random seeds.

Figure 8.7 shows the sorted best objective function values in 1000 evaluations (left plot) or 500 evaluations (right plot). Within 1000 evaluations, the RCDS method with the initial conjugate direction set reached the loss rate below -2.5 mA for most cases. The corresponding coupling ratio is about 0.025%. In many cases, the RCDS method has reached the -2.5 mA level within 500 evaluations. RCDS with the unit vectors initially ("RCDS-Imat") is not as efficient as with the initial conjugate direction set; but it still outperforms the simplex methods. The RSimplex method does better than the N-M simplex method with 1000 evaluations; but within 500 evaluations the two have similar performances.

Figure 8.8 shows the convergence histories of the cases with the 30th final minimum value for the four methods. The objective function and the first 8 skew quadrupoles are shown. It can be seen that the N-M simplex method has a fast convergence initially, but stopped making any further gains after reaching the -1.5 mA/min level as the simplex has shrunk to a tiny size. The other three methods continue to explore the parameter space and make gains throughout the runs.

8.3.2 Experiments

Experimental tests of vertical emittance minimization have been done on SPEAR3 [57]. The knobs are the same 13 skew quadrupoles as used in simulation. The initial solution is with all knobs turned off. The corresponding

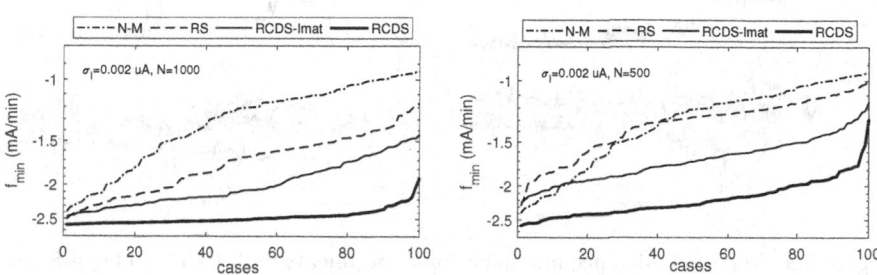

Figure 8.7 The final loss rate achieved by the four methods: N-M simplex, RSimplex, RCDS with unit vectors for the initial direction set (RCDS-Imat), and RCDS with calculated conjugate direction set (RCDS), in simulation for 100 cases each. Loss rate noise sigma is 0.03 mA/min (not included in the plots). Left: with 1000 evaluations; right: with 500 evaluations.

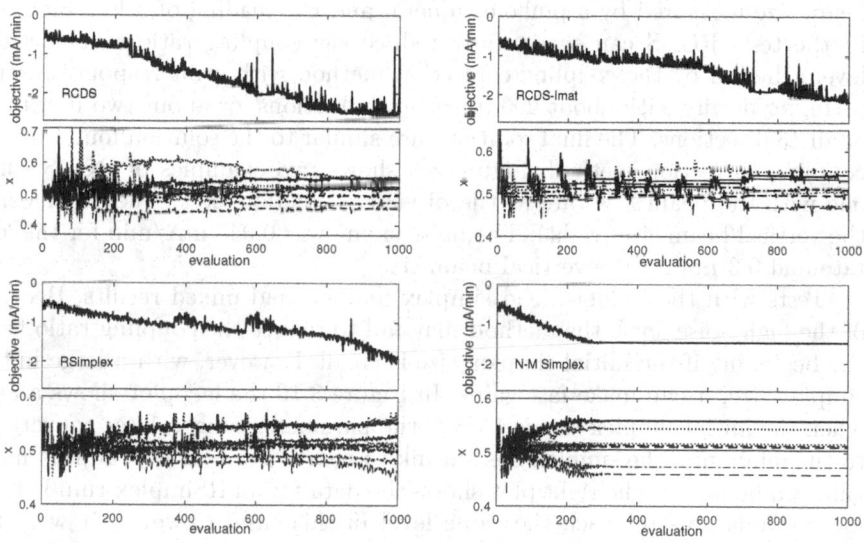

Figure 8.8 The convergence history in simulation for the case with a final minimum ranking the 30th (out of 100 runs) for the four methods: RCDS with conjugate direction set (top left), RCDS with unit vectors (top right), RSimplex (bottom left), and N-M simplex (bottom right).

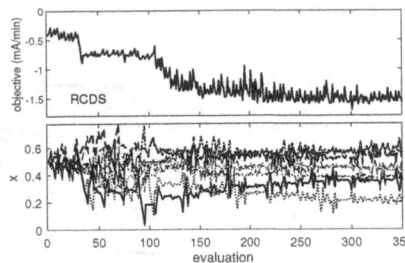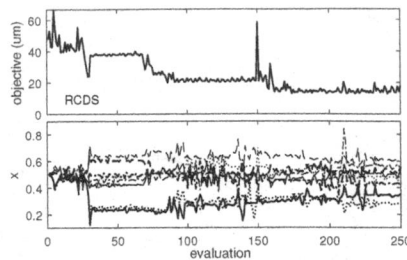

Figure 8.9 Vertical emittance minimization experiments with RCDS. The objective functions are (negative) the beam loss over 6 seconds (left) or the photon beam size σ_y measured by the pinhole camera (right).

emittance ratio as determined by fitting the orbit response matrix (with 42 skew quadrupoles as fitting parameters) is 1.1%. The parameter range for the skew quadrupoles is $[-20, 20]$ A in power supply setpoints. After a new solution is dialed in, the program waits for 2 seconds for the magnets to settle to the new setpoints.

Multiple choices of performance measures have been used for the test of the RCDS algorithm, including the beam loss over a 6-second period, the vertical beam size measured by a pinhole camera, and the reading of a loss monitor. In the tests RCDS can successfully reduce the coupling ratio to below the level achieved by the coupling correction method with orbit response matrix fitting, typically with about 200 function evaluations, or about two iterations of all 13 directions. The final solutions are similar to the solution found by the coupling correction method. Figure 8.9 shows two examples of RCDS runs, one with the beam loss rate as the objective function (left), the other using the vertical beam size (right). The noise sigma was 0.045 mA/min for the loss rate and 0.3 μm for the vertical beam size.

Tests with the Nelder-Mead simplex method had mixed results. Because of the high noise level, the method may fail to reduce the coupling ratio from the beginning if the initial simplex size is small. However, with a large initial simplex size, it can make fast gains. In Figure 8.10 the left plot shows a case when the initial simplex size is 15% of the parameter range. After converging to the minimum, the simplex has shrunk to nearly a size of zero and no more gain can be made. The right plot shows the data for an RSimplex run. It took more evaluations to reach the same level in this case. However, it was still reducing the vertical emittance at the end.

The particle swarm optimization algorithm has also been tested on the vertical emittance minimization problem, using the loss monitor reading as the objective. The loss monitor measurement is much faster than waiting for a sizable beam charge decrease. The algorithm converged to the minimum with about 3000 function evaluations [54]. As a comparison, using the same setup, the NSGA-II method took 20,000 evaluations to reach the same coupling level [115].

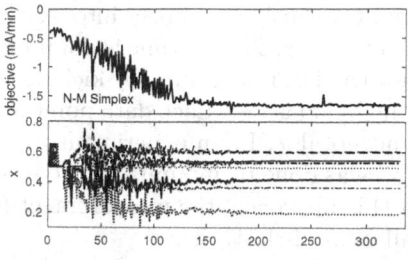 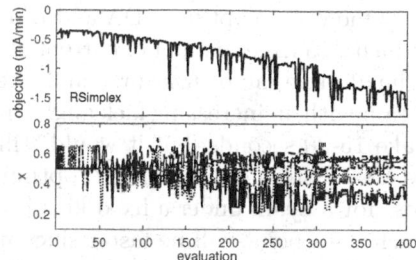

Figure 8.10   Vertical emittance minimization experiments on SPEAR3 with the N-M simplex (left) and the RSimplex (right) methods.

8.4   OPTIMIZATION OF NONLINEAR BEAM DYNAMICS

The operation of a storage ring requires a large dynamic aperture (DA) for a high injection efficiency and large local momentum apertures (LMA) for a high Touschek lifetime. For low emittance storage rings, the strong nonlinear fields of sextupole magnets in the lattice can severely limit the DA and the LMA. A storage ring lattice design relies on a delicate scheme for the cancellation of the contributions of the sextupoles to the various nonlinear resonances. The sextupole scheme is sensitive to the magnetic field errors in the lattice; the DA and LMA performance typically degrade when errors are introduced to the machine. Since a real machine will always differ from the design lattice, even after linear optics correction, the nonlinear beam dynamics performance of an actual storage ring usually is not as good as the design performance.

The effects of the field errors can be compensated with knobs that affect the nonlinear beam dynamics of the beams, e.g., the sextupole strengths. Beam-based correction of nonlinear beam dynamics has been discussed in Chapter 6. As pointed out there, beam-based correction for nonlinear dynamics is very difficult as the RDT signals are very weak and there is no clear connection between the signals and the DA and LMA. Beam-based optimization of the nonlinear beam dynamics performance has been proven to be an effective method. With this approach, the nonlinear dynamics knobs are varied by optimization algorithms to directly improve the injection efficiency or the Touschek lifetime, which automatically leads to a better DA or LMA.

Optimization of the DA and the LMA has been demonstrated experimentally [60, 91, 123, 81]. While it is desirable to optimize DA and LMA simultaneously, so far in experiments the two are optimized separately. This is because during a DA optimization the beam current changes due to injection or beam loss, while the LMA optimization prefers a steady beam current. However, in practice this has not been a problem since a lattice with a good DA often also has good LMAs and vice versa, even though there exist lattices that are good in one of the two metrics but not the other.

One way to optimize DA and LMA simultaneously is to inject into a single bunch to a fixed amount of current at which the Touschek lifetime is dominant. The lifetime can be measured and after that the DA is measured by kicking the beam with an increasing kick angle, until the beam is lost. Each data point may take 15-30 seconds, but it would still be acceptable. In an experiment at the MAX-IV storage ring, the DA optimization was done by minimizing the beam loss for a beam under a fixed kick [91]. In this approach, the measurement for each data point will be faster since not all beam is lost.

In the next we will discuss the DA and LMA optimization separately.

8.4.1 Dynamic aperture optimization

General considerations: As mentioned in the above, the DA can be measured by kicking a low charge beam with an increasing kick strength and recording the kick strength when the beam loss occurs. The measured DA can then be used as the objective function. The beam charge can be as low as $\sim 10 \ \mu A$, which helps limit the accumulated beam loss during the experiment. This approach would be particularly useful for the commissioning of a new ring before high current beams can be stored.

An alternative approach is to optimize the injection efficiency directly. If the tuning knobs change only the DA and not the other factors in the injection process that affect the injection efficiency, optimizing the injection efficiency is equivalent to optimizing the DA. Since the injection efficiency cannot go above 100% or below 0%, if the DA is so small such that no beam is captured, or if the DA is so large such that the injection efficiency is already 100%, the optimization algorithms could fail since the objective function may not respond to the tuning knob changes. In such cases, changes may be made to bring the injection efficiency to the medium range. If the DA is initially too small, random settings of the tuning knobs may be tried until one solution is found to allow some injected beam to be captured. If the DA is large enough to have a high initial injection efficiency, yet it is still desired to further increase the DA, the injected beam can be mis-steered from the ideal trajectory to lower the injection efficiency. In the case of off-axis injection, which is the injection scheme adopted for the existing storage rings, the injection efficiency can be lowered by reducing the kicker bump.

Figure 8.11 illustrates the horizontal off-axis injection of a storage ring. The centers of the stored beam and the injected beam are normally separated by a distance D. During injection, the stored beam is kicked toward the septum wall. To fully capture the injected beam, the DA needs to be larger than the distance between the injected beam and the kicked stored beam plus half of the size of the injected beam. If the DA is smaller, the part of the injected beam beyond the DA will be lost. If the kicker bump is reduced, the DA boundary will shift with the stored beam and the loss of the injected beam will occur if the boundary moves into the injected beam. For the injection

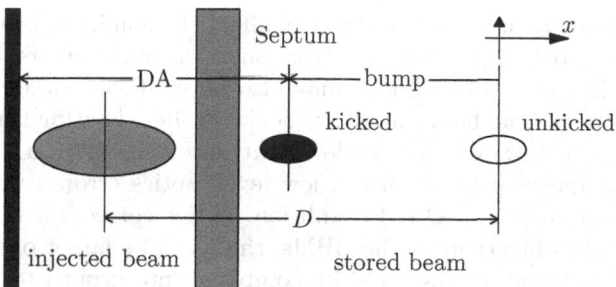

Figure 8.11 Illustration of off-axis injection for a storage ring.

efficiency to have the highest sensitivity over the DA, the boundary should better be in the dense part of the beam.

The tuning knobs for the DA are usually the power supply setpoints for the sextupole magnets. Sextupoles are the sources of the nonlinearity in the beam motion and are hence effective knobs to change the DA. The purpose of including sextupoles in the lattice is to correct the chromaticities. Any change to the sextupole setting should not significantly change the chromaticity values. Some storage rings have harmonic sextupoles, which are located in areas with a zero dispersion. The harmonic sextupoles do not change the chromaticities and are free tuning knobs for the nonlinear beam dynamics. More often is the case in which all sextupoles are at dispersive locations; in some cases harmonic sextupoles become chromatic sextupoles after the linear lattices are modified for emittance reduction. In such cases, care should be taken to not change the chromaticities during the DA optimization.

There are usually a small number of sextupoles in each lattice cell placed symmetrically about the cell center. Sextupoles at the symmetric points of the same cell and sextupoles at the same location of different cells may be powered by a serial power supply. Sextupoles on the same power supply are collectively referred to as a sextupole family. At the minimum, there are only two sextupole families, which are required for chromaticity correction and leave no free knobs for DA optimization. In the other extreme, all sextupoles could be powered individually. There would be tens or hundreds of sextupole knobs, potentially more than the optimization algorithms can handle in online applications. In such a case, the sextupoles need to be grouped for the number of knobs to be reduced to an acceptable level (e.g., 10 to 30). Typically the sextupoles are grouped such that the each family consists of a number of magnets that are symmetrically distributed around the ring. In some cases, the sextupoles may be already grouped and powered in several families according to their locations. Some storage rings have octupole magnets in the lattice for controlling the nonlinear dynamics behavior [91]. They should also be included as nonlinear dynamics tuning knobs.

Since a major path for lattice errors to affect the nonlinear beam dynamics performance is by introducing betatron phase advance errors between the sextupoles, the linear optics knobs may also be used for tuning the DA and LMA. Linear optics can be controlled to below 1% beta beating through beam-based correction. However, for the low emittance rings with a large number of strong sextupoles, even at such a low level, optics errors can still have a significant impact on the DA. In addition, optics correction normally only targets the optics functions at the BPMs; there can be larger optics errors at the sextupoles. Quadrupoles could be combined into groups to form tuning knobs that affect the phase advances on the sextupoles with various patterns, using the phase advance response matrix with respect to the quadrupoles.

The SPEAR3 DA optimization experiment: In the beam-based DA optimization for the SPEAR3 storage ring [60], the injection efficiency was used as the objective function and the injection kicker bump was reduced. The SPEAR3 storage ring has 18 double-bend achromat (DBA) cells, 14 of which are standard cells and the remaining 4 are matching cells. There are two pairs of sextupoles in each cell. Originally SPEAR3 had only 4 sextupole families: the SF/SD families for the standard cells and the SFM/SDM families for the matching cells. For the purpose of optimizing DA/LMA for a lattice upgrade, the SF/SD families were split into 4 families each, 3 of which consist of SF (SD) magnets in 4 cells symmetrically distributed around the ring, while the last one only consists of sextupoles in the two cells at the centers of the two arcs. The sextupole family distribution in half of the ring may be labeled M-X_2-X_3-X_4-X_5-X_4-X_3-X_2-M, where M stands for the matching cells and X_i for the standard cells. The other half is in the mirror symmetry. For example, all the SF sextupoles in X_2 cells make one family. There are a total of 10 sextupole families.

To be able to freely change the sextupole settings without changing the chromaticities, combined knobs were created by applying singular value decomposition (SVD) to the chromaticity response matrix, $\mathbf{R}_{\mathrm{chrom}} = \mathbf{USV}^T$. The chromaticities, $(C_x, C_y)^T$, depend on the sextupole families through linear relationships. The derivatives of the chromaticities with respect to the sextupole families constitute the chromaticity response matrix, which is a 2×10 matrix for the SPEAR3 case. It has two non-zero singular values, which corresponds to the sextupole combinations that can change chromaticities. The other 8 singular values are zeros and the corresponding column vectors in $\mathbf{V}$ form the basis vectors for the null space; any sextupole changes in the null space do not change the chromaticities. The eight null space basis vectors are used as tuning knobs in the SPEAR3 DA optimization experiments. Some SVD modes of the SPEAR3 chromaticity response matrix are shown in Figure 8.12.

The injection efficiency was measured through the beam current change in the storage ring and the intensity of the injected beam. Each data point was collected with injection for 10 seconds to reduce the noise level. The noise sigma for injection efficiency is about 3%. Figure 8.13 shows the evolution of the injection efficiency and 4 out of the 10 sextupole families for an

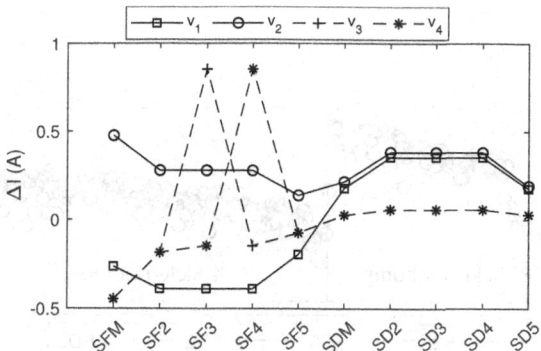

Figure 8.12   Selected SVD modes of the SPEAR3 chromaticity response matrix. v_1 and v_1 are the two modes with non-zero SVs. v_3 and v_4 are two out of the eight modes with zero SVs, which are used as DA tuning knobs.

optimization run with RCDS, starting with the solution when all standard cell SF and SD families are at equal strengths, respectively. Because SPEAR3 normally had good injection efficiency, the kicker bump was reduced to 85% of the nominal size initially for the injection efficiency to drop to about 35%. During the optimization, the injection efficiency approached 100% after about 90 evaluations. To make room for more gains in the DA, the kicker bump was reduced further to 77% and the RCDS was re-launched. With a total of 230 evaluations, the DA was significantly improved. The kick angle for kicker K1 required to kick out the beam was increased from 1.55 mrad to 2.13 rad after the optimization, corresponding to a dynamic aperture of 15.1 mm and 20.6 mm, respectively. The performance of the optimized solution was also verified by measuring injection efficiency with a reduced kicker bump. As shown in the left plot of Figure 8.14, full capture of the injected beam can be achieved with a kicker bump that is smaller than the nominal sextupole solution by about 5 mm. In Figure 8.14 the right plots compare the currents of the sextupoles before and after the optimization.

The chromaticities were measured before and after the optimization and were at +3 for both planes with little changes. The Touschek lifetime of the beam was characterized for the initial and optimized solutions using beam lifetime vs. RF voltage scans [111]. In both cases, the Touschek lifetime was limited by the RF bucket height and the measured beam lifetime was equal when the coupling ratio was equal. Therefore, there was no negative impact on the beam lifetime.

The particle swarm optimization (PSO) method was also applied to DA optimization of SPEAR3. Starting with an intermediate solution found by RCDS for which the DA improved by about 2 mm from the original lattice, it was able to find a lattice solution similar to the final RCDS solution in about 300 evaluations.

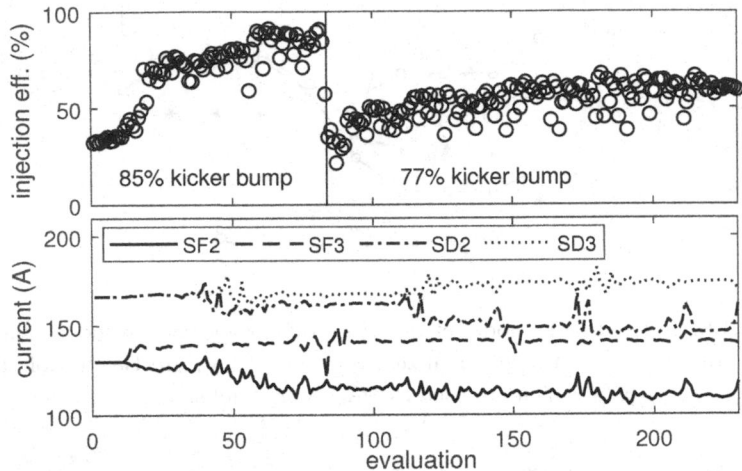

Figure 8.13 Optimization of the DA for SPEAR3 with the RCDS method. Top: injection efficiency; bottom: the currents of 4 out of 10 sextupole families. The kicker bump was first reduced to 85% of the nominal size, then to 77% after the DA was improved. Modified from Ref. [60] under the Creative Commons Attribution 3.0 License.

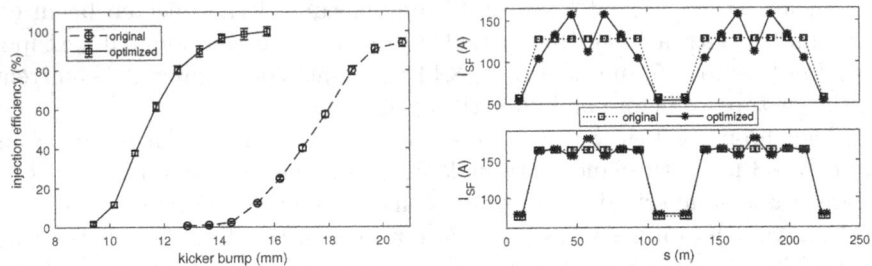

Figure 8.14 Comparison of the lattice solutions before and after the DA optimization with RCDS for SPEAR3. Left: injection efficiency vs. the kicker bump amplitude; right: the sextupole power supply setpoints. Modified from Ref. [60] under the Creative Commons Attribution 3.0 License.

DA optimization at MAX-IV and NSLS-II: DA optimization with RCDS was also successfully applied on the MAX-IV storage ring [91]. MAX-IV is the first storage ring light source that adopted an multi-bend achromat (MBA) lattice. It currently has the lowest emittance among all storage ring light sources. In the MAX-IV experiments, the objective function was the fractional beam loss of the stored beam under a fixed horizontal kick. The tuning knobs are five sextupole families and three octupole families. The chromaticity response matrix of the sextupoles was used to construct three combined knobs that do not change chromaticities. There were a total of 6 free knobs. RCDS steadily improved the DA over 4 iterations. The DA was increased roughly from 5 mm to 7 mm (measured at the center of a long straight section), while the chromaticities remained about the same. The optimized lattice also has an significantly improved momentum acceptance, which reached the design goal of 4.5%.

On the NSLS-II storage ring, DA optimization with RCDS using sextupole knobs improved the DA by more than 20% [123]. It was found that the horizontal tune shift with amplitude was reduced by more than a factor of two in the optimized lattice.

8.4.2 Local momentum aperture optimization

If the local momentum aperture (LMA) is smaller than the RF bucket height, it becomes the limiting factor for the Touschek lifetime. In such cases, online optimization may be done to enlarge the LMA. The tuning knobs for the LMA are usually the same as those for the DA optimization. The objective function can be the beam loss rate for a Touschek scattering dominated beam. The Touschek loss rate is strongly dependent on the momentum aperture, δ_A, with $\frac{1}{\tau_T}$ roughly proportional to δ_A^{-3}. As shown in Eq. 8.2, it is also proportional to the charge density. In the LMA optimization experiment, it is critical to separate the charge density factor out from the loss rate in order to target only the LMA.

In an experiment, the bunch charge constantly changes as a consequence of the beam loss that causes a finite lifetime. In addition, the bunch length varies with the bunch charge due to bunch lengthening through the longitudinal impedance. To avoid a significant impact from bunch lengthening, the bunch charge may be replenished frequently so that it is roughly constant. When sextupole strengths are changed, the vertical emittance could also be altered, as vertical orbit offsets in sextupoles are a source of linear coupling errors. To mitigate the impact of vertical emittance variation, it is desirable to keep a constant orbit (with orbit feedback) that is as close to the sextupole centers as possible. It would also be helpful to slightly increase the coupling ratio with skew quadrupoles, such that coupling variations due to sextupoles are small compared to the initial vertical emittance. If the diagnostics are available, a preferred approach is to measure the vertical emittance and factor out its effect from the objective function.

Beam-based optimization of LMA was first demonstrated on the ESRF storage ring [81]. In the experiment, the objective function was defined as

$$f(\mathbf{x}) = -\tau(\mathbf{x})\frac{I\sigma_{z0}\sigma_{y,0}}{I_0\sigma_z(I)\sigma_y},$$ (8.5)

where $\sigma_z(I)$ is the bunch length calculated with an impedance model, $\sigma_{z0} = \sigma_z(I_0)$, σ_y is the average vertical beam size for the measurements at 13 locations distributed around the ring. Care was also taken to wait for at least 30 minutes after each injection of a fresh beam injection to allow the beam spin polarization distribution to settle in an equilibrium state as it could affect the Touschek lifetime by up to 15%. With the beam density factor separated out from the objective function and spin polarization effect stabilized, the optimization targets only the LMA.

Two sets of sextupole knobs were used in the experiments. In one setup, 12 sextupole correctors were used for a fill pattern with many low-charge bunches. The lifetime was increased from 47.5 hr to 75.5 hr by RCDS with about 5 iterations in 40 min. In the other setup, 10 sextupole knobs that consisted of 5 main sextupole families and additional sextupoles with individual power supplies were used on a fill pattern with 16 high-charge bunches. The lifetime was increased from 11 hr to 17 hr. The optimized lattice had been used in operation.

Bibliography

[1] M. Aiba, M. Böge, J. Chrin, N. Milas, T. Schilcher, and A. Streun. Comparison of linear optics measurement and correction methods at the Swiss Light Source. *Phys. Rev. ST Accel. Beams*, 16:012802, Jan 2013.

[2] M. Aiba, M. Böge, N. Milas, and A. Streun. Ultra low vertical emittance at SLS through systematic and random optimization. *Nucl. Instrum. Methods Phys. Res, Section A*, 694:133 – 139, 2012.

[3] M. Aiba, S. Fartoukh, A. Franchi, M. Giovannozzi, V. Kain, M. Lamont, R. Tomas, G. Vanbavinckhove, J. Wenninger, F. Zimmermann, R. Calaga, and A. Morita. First β-beating measurement and optics analysis for the CERN Large Hadron Collider. *Phys. Rev. ST Accel. Beams*, 12:081002, Aug 2009.

[4] Peter Auer. Using confidence bounds for exploitation-exploration trade-offs. *Journal of Machine Learning Research*, 3:397–422, 2002.

[5] B. Autin and Y. Marti. Closed orbit correction of A.G. machines using a small number of magnets, 1973.

[6] R. Bartolini. Resonance driving term experiments: an overview. In *Proceedings of ICAP'06*, pages 22–27, 2006.

[7] R. Bartolini, M. Giovannozzi, W. Scandale, A. Bazzani, and E. Todesco. Algorithms for a precise determination of the betatron tune. In *Proceedings of EPAC'96*, 1996.

[8] R. Bartolini, I. P. S. Martin, G. Rehm, and F. Schmidt. Calibration of the nonlinear ring model at the Diamond light source. *Phys. Rev. ST Accel. Beams*, 14:054003, May 2011.

[9] R. Bartolini, I. P. S. Martin, J. H. Rowland, P. Kuske, and F. Schmidt. Correction of multiple nonlinear resonances in storage rings. *Phys. Rev. ST Accel. Beams*, 11:104002, Oct 2008.

[10] R. Bartolini and F. Schmidt. Normal form via tracking or beam data. *Part. Accel.*, 59:93–106, December 1997.

[11] A. Bazzani et al. Normal forms for Hamiltonian maps and nonlinear effects in a particle accelerator. *Nuovo Cim. B*, 102:51–80, 1988.

[12] A. Bazzani, E. Todesco, G. Turchetti, and G. Servizi. A normal form approach to the theory of nonlinear betatronic motion. CERN 94-02, 1994.

[13] A. Belouchrani, K. Abed-Meraim, J. Cardoso, and E. Moulines. A blind source separation technique using second-order statistics. *IEEE Transactions on Signal Processing*, 45(2):434–444, Feb 1997.

[14] W. F. Bergan, I. V. Bazarov, C. J. R. Duncan, D. B. Liarte, D. L. Rubin, and J. P. Sethna. Online storage ring optimization using dimension-reduction and genetic algorithms. *Phys. Rev. Accel. Beams*, 22:054601, May 2019.

[15] J. Borer, A. Hofmann, J. Koutchouk, T. Risselada, and B. Zotter. Measurements of betatron phase advance and beta function in the ISR. *IEEE Transactions on Nuclear Science*, 30(4):2406–2408, Aug 1983.

[16] M. Borland, V. Sajaev, L. Emery, and A. Xiao. Direct methods of optimization of storage ring dynamic and momentum aperture. In *Proceedings of PAC'09*, pages 3850–3852, 2009.

[17] Karl L. Brown. A first-and second-order matrix theory for the design of beam transport systems and charged particle spectrometers. SLAC-75 and SLAC-PUB-3381, 1972.

[18] J. Cardoso and A. Souloumiac. Jacobi angles for simultaneous diagonalization. *SIAM Journal on Matrix Analysis and Applications*, 17(1):161–164, 1996.

[19] P. Castro, J. Borer, A. Burns, G. Morpurgo, and R. Schmidt. Betatron function measurement at LEP using the BOM 1000 turns facility. In *Proceedings of PAC'93*, pages 2103–2105, 1993.

[20] S. Choi, A. Cichocki, H.M. Park, and S.Y. Lee. Blind source separation and independent component analysis: A review. *Neural Information Processing - Letters and Reviews*, 6(1):1–57, Jan. 2005.

[21] Y. Chung, G. Decker, and K. Evans. Closed orbit correction using singular value decomposition of the response matrix. In *Proceedings of PAC'93*, pages 2263–2265, 1993.

[22] Y. Chung, G. Decker, and K. Evans. Measurement of beta-function and phase using the response matrix. In *In Proceedings of PAC'93*, pages 188–190, 1993.

[23] Y. Chung, J. Kirchman, F. Lenkszus, A. J. Votaw, R. Hettel, W. J. Corbett, D. Keeley, J. Sebek, C. Wermelskirchen, and J. Yang. Closed

orbit feedback with digital signal processing. In *Proceedings of EPAC'94*, pages 1592–1595, 1994.

[24] Pierre Comon. Independent component analysis, a new concept? *Signal Processing*, 36(3):287–314, 1994.

[25] W.J. Corbett, M.J. Lee, and V. Ziemann. A fast model-calibration procedure for storage rings. In *Proceedings of PAC'93*, pages 108–110, 1993.

[26] E.D Courant and H.S Snyder. Theory of the alternating-gradient synchrotron. *Annals of Physics*, 3(1):1–48, 1958.

[27] K. Deb, A. Pratap, S. Agarwal, and T. Meyarivan. A fast and elitist multiobjective genetic algorithm: NSGA-II. *Trans. Evol. Comp*, 6(2):182–197, April 2002.

[28] Kalyanmoy Deb. *Multi-Objective Optimization Using Evolutionary Algorithms*. Wiley, New York, 2001.

[29] Kalyanmoy Deb and Hans-georg Beyer. Self-adaptive genetic algorithms with simulated binary crossover. *Evol. Comput.*, 9(2):197–221, June 2001.

[30] R. Dowd, Y-R. E. Tan, and K. P. Wootton. Vertical emittance at the quantum limit. In *Proceedings of IPAC'14*, pages 1096–1098, 2014.

[31] Alex J. Dragt and John M. Finn. Normal form for mirror machine Hamiltonians. *Journ. of Math. Phys.*, 20:2649–2660, December 1979.

[32] L. Emery, M. Borland, and H. Shang. Use of a general-purpose optimization module in accelerator control. In *Proceedings of the PAC'03*, pages 2330–2332, 2003.

[33] P. Emma and W. Spencer. Grid scans: A transfer map diagnostic. In *Proceedings of PAC'91*, pages 1549–1551, 1991.

[34] W. Fischer, J. Beebe-Wang, Y. Luo, S. Nemesure, and L. K. Rajulapati. RHIC proton beam lifetime increase with 10- and 12-pole correctors. In *Proceedings of the IPAC'10*, pages 4752–4754, 2010.

[35] J. W. Flanagan, K. Oide, N. Akasaka, A. Enomoto, K. Furukawa, T. Kamitani, H. Koiso, Y. Ogawa, S. Ohsawa, and T. Suwada. A simple real-time beam tuning program for the KEKB injector linac. In *Proceedings of the 1998 International Computational Accelerator Physics Conference, Monterey, CA*, 1998.

[36] E. Forest. A Hamiltonian free description of single particle dynamics for hopelessly complex periodic systems. *Journ. of Math. Phys.*, 31:1133–1144, June 1990.

[37] Etienne Forest. *Beam Dynamics: A New Attitude and Framework*. Harwood Academic, 1998.

[38] Etienne Forest and Ronald D. Ruth. Fourth-order symplectic integration. *Physica D: Nonlinear Phenomena*, 43(1):105 – 117, 1990.

[39] A. Franchi, L. Farvacque, J. Chavanne, F. Ewald, B. Nash, K. Scheidt, and R. Tomas. Vertical emittance reduction and preservation in electron storage rings via resonance driving terms correction. *Phys. Rev. ST Accel. Beams*, 14:034002, Mar 2011.

[40] A. Franchi, L. Farvacque, F. Ewald, G. Le Bec, and K. B. Scheidt. First simultaneous measurement of sextupolar and octupolar resonance driving terms in a circular accelerator from turn-by-turn beam position monitor data. *Phys. Rev. ST Accel. Beams*, 17:074001, Jul 2014.

[41] G. Golub. Numerical methods for solving linear least squares problems. *Numerische Mathematik*, 7(3):206–216, Jun 1965.

[42] G. H. Golub and C. Reinsch. Singular value decomposition and least squares solutions. *Numerische Mathematik*, 14(5):403–420, Apr 1970.

[43] W. Guo, S. Kramer, F. Willeke, X. Yang, and L. Yu. A lattice correction approach through betatron phase advance. In *Proceedings of IPAC'16*, pages 62–65, 2016.

[44] M. Harrison and S. Peggs. Global beta measurement from two perturbed closed orbits. In *Proceedings of PAC'87*, pages 1105–1107, 1987.

[45] L. M. Healy. Lie algebraic methods for treating lattice parameter errors in particle accelerators. Doctoral thesis, University of Maryland, unpublished, 1986.

[46] R. O. Hettel. Beam Steering at the Stanford Synchrotron Radiation Laboratory. *IEEE Transactions on Nuclear Science*, 30:2228, August 1983.

[47] Y. Hidaka, B. Podobedov, and J. Bengtsson. Linear optics characterization and correction method using turn-by-turn BPM data based on resonance driving terms with simultaneous BPM calibration capability. In *Proceedings of NAPAC'16*, pages 605–608, 2016.

[48] A. Hofmann and B. Zotter. Measurement of the β-functions in the ISR. ISR-TH-AH-BZ-amb, CM-P00072144, 1975.

[49] X. Huang, S. Y. Lee, E. Prebys, and C. Ankenbrandt. Fitting the fully coupled ORM for the Fermilab Booster. In *Proceedings of PAC'05*, pages 3322–3324, 2005.

[50] X. Huang, J. Safranek, and G. Portmann. LOCO with constraints and improved fitting technique. *ICFA Newsletter*, 44:60–69, 2007.

[51] X. Huang and X. Yang. Linear optics and coupling correction with turn-by-turn BPM data. In *Proceedings of IPAC'15*, pages 698–701, 2015.

[52] Xiaobiao Huang. Beam diagnosis and lattice modeling of the Fermilab Booster. Doctoral thesis, Indiana University, unpublished, 2005.

[53] Xiaobiao Huang. Matrix formalism of synchrobetatron coupling. *Phys. Rev. ST Accel. Beams*, 10:014002, Jan 2007.

[54] Xiaobiao Huang. Development and application of online optimization algorithms. In *Proceedings of NAPAC2016*, pages 1287–1291, Chicago, IL, 2016.

[55] Xiaobiao Huang. Robust simplex algorithm for online optimization. *Phys. Rev. Accel. Beams*, 21:104601, Oct 2018.

[56] Xiaobiao Huang. Multi-objective multi-generation Gaussian process optimizer for design optimization. *ArXiv*, abs/1907.00250, 2019.

[57] Xiaobiao Huang, Jeff Corbett, James Safranek, and Juhao Wu. An algorithm for online optimization of accelerators. *Nucl. Instrum. Methods Phys. Res, Section A*, 726:77 – 83, 2013.

[58] Xiaobiao Huang, S. Y. Lee, Eric Prebys, and Ray Tomlin. Application of independent component analysis to Fermilab Booster. *Phys. Rev. ST Accel. Beams*, 8:064001, Jun 2005.

[59] Xiaobiao Huang and James Safranek. Nonlinear dynamics optimization with particle swarm and genetic algorithms for SPEAR3 emittance upgrade. *Nucl. Instrum. Methods Phys. Res, Section A*, 757:48–53, 09 2014.

[60] Xiaobiao Huang and James Safranek. Online optimization of storage ring nonlinear beam dynamics. *Phys. Rev. ST Accel. Beams*, 18:084001, Aug 2015.

[61] Xiaobiao Huang, Jim Sebek, and Don Martin. Lattice modeling and calibration with turn-by-turn orbit data. *Phys. Rev. ST Accel. Beams*, 13:114002, Nov 2010.

[62] A. Hyvarinen. Fast and robust fixed-point algorithms for independent component analysis. *IEEE Transactions on Neural Networks*, 10(3):626–634, May 1999.

[63] A. Hyvarinen, J. Karhunen, and E. Oja. *Independent Component Analysis*. Wiley, New York, 2001.

[64] A. Hyvärinen and E. Oja. Independent component analysis: algorithms and applications. *Neural Networks*, 13(4):411 – 430, 2000.

[65] J. Irwin, C. X. Wang, Y. T. Yan, K. L. F. Bane, Y. Cai, F.-J. Decker, M. G. Minty, G. V. Stupakov, and F. Zimmermann. Model-independent beam dynamics analysis. *Phys. Rev. Lett.*, 82:1684–1687, Feb 1999.

[66] J. Irwin and C. x. Wang. Explicit soft fringe maps of a quadrupole. In *Proceedings of PAC'95*, pages 2376–2378, 1995.

[67] F. C. Iselin. Lie transformations and transport equations for combined-function dipoles. *Part. Accel.*, 17:143–155, March 1985.

[68] John David Jackson. *Classical Electrodynamics*. Wiley, New York, 3rd ed., 1999.

[69] Donald R. Jones, Matthias Schonlau, and William J. Welch. Efficient global optimization of expensive black-box functions. *Journal of Global Optimization*, 13(4):455–492, Dec 1998.

[70] Christian Jutten and Jeanny Herault. Blind separation of sources, part I: An adaptive algorithm based on neuromimetic architecture. *Signal Processing*, 24(1):1 – 10, 1991.

[71] J. Kennedy and R. Eberhart. Particle swarm optimization. In *Proceedings of ICNN'95 - International Conference on Neural Networks*, volume 4, pages 1942–1948 vol.4, Nov 1995.

[72] H. J. Kushner. A new method of locating the maximum point of an arbitrary multipeak curve in the presence of noise. *Journal of Basic Engineering*, 86(1):97, 1964.

[73] A. Langner, G. Benedetti, M. Carla, U. Iriso, Z. Marti, J. Coello de Portugal, and R. Tomas. Utilizing the N beam position monitor method for turn-by-turn optics measurements. *Phys. Rev. Accel. Beams*, 19:092803, Sep 2016.

[74] J. Laskar. Frequency map analysis and particle accelerators. In *Proceedings of PAC'03*, pages 378–382, 2003.

[75] Jacques Laskar. Frequency analysis for multi-dimensional systems. Global dynamics and diffusion. *Physica D: Nonlinear Phenomena*, 67(1):257 – 281, 1993.

[76] M. Lee, S. Kleban, S. Clearwater, W. Scandale, T. Pettersson, H. Kugler, A. Riche, M. Chanel, and E. Martensson. SLAC-PUB-4411, 1987.

[77] M. Lee, J. Sheppard, M. Sullenberger, and M. Woodley. SLAC-PUB-3217, 1983.

[78] S. Y. Lee. *Accelerator Physics*. World Scientific, 1999.

[79] Kenneth Levenberg. A method for the solution of certain non-linear problems in least squares. *Quart. Appl. Math.*, 2:64–168, 1944.

[80] Robert Michael Lewis, Virginia Torczon, and Michael W. Trosset. Direct search methods: then and now. *Journal of Computational and Applied Mathematics*, 124(1):191 – 207, 2000. Numerical Analysis 2000. Vol. IV: Optimization and Nonlinear Equations.

[81] S.M. Liuzzo, N. Carmignani, L. Farvacque, Perron T. Nash, B., P. Raimondi, R. Versteegen, and S. M. White. RCDS optimizations for the ESRF storage ring. In *Proceedings of IPAC2016*, pages 3420–3422, Busan, Korea, 2016.

[82] D. Lizotte. *Practical Bayesian Optimization*. PhD thesis, University of Alberta, Edmonton, Alberta, Canada, 2008.

[83] Yun Luo. Linear coupling parametrization in the action-angle frame. *Phys. Rev. ST Accel. Beams*, 7:124001, Dec 2004.

[84] L. Malina, J. Coello de Portugal, T. Persson, P. K. Skowronski, R. Tomas, A. Franchi, and S. Liuzzo. Improving the precision of linear optics measurements based on turn-by-turn beam position monitor data after a pulsed excitation in lepton storage rings. *Phys. Rev. Accel. Beams*, 20:082802, Aug 2017.

[85] D. Marquardt. An algorithm for least-squares estimation of nonlinear parameters. *Journal of the Society for Industrial and Applied Mathematics*, 11(2):431–441, 1963.

[86] J Mockus, Vytautas Tiesis, and Antanas Zilinskas. *The Application of Bayesian Methods for Seeking the Extremum*, volume 2, pages 117–129. Elsevier, 1978.

[87] Jonas Mockus. On Bayesian methods for seeking the extremum and their application. In *IFIP Congress*, pages 195–200, 1977.

[88] Laurent S. Nadolski. LOCO fitting challenges and results for SOLEIL. *ICFA Newsletter*, 44:69–81, 2007.

[89] J. A. Nelder and R. Mead. A simplex method for function minimization. *The Computer Journal*, 7(4):308–313, 01 1965.

[90] Kazuhito Ohmi, Kohji Hirata, and Katsunobu Oide. From the beam-envelope matrix to synchrotron-radiation integrals. *Phys. Rev. E*, 49:751–765, Jan 1994.

[91] D. K. Olsson. Online optimisation of the MAX-IV 3 GeV ring dynamic aperture. In *Proceedings of IPAC2018*, pages 2281–2283, Vancouver, BC, Canada, 2018.

[92] X. Pang and L.J. Rybarcyk. Multi-objective particle swarm and genetic algorithm for the optimization of the LANSCE linac operation. *Nucl. Instrum. Methods Phys. Res, Section A*, 741:124 – 129, 2014.

[93] G. Portmann, J. Corbett, and A. Terebilo. An accelerator control middle layer using Matlab. In *Proceedings of PAC'05*, pages 4009–4011, 2005.

[94] G. Portmann, D. Robin, and L. Schachinger. Automated beam based alignment of the ALS quadrupoles. In *Proceedings of PAC'95*, pages 2693–2695, 1995.

[95] Greg Portmann, J. Safranek, and X. Huang. Matlab based LOCO. *ICFA Newsletter*, 44:49–60, 2007.

[96] M. J. D. Powell. An efficient method for finding the minimum of a function of several variables without calculating derivatives. *The Computer Journal*, 7(2):155–162, 1964.

[97] William H. Press, Saul A. Teukolsky, William T. Vetterling, and Brian P. Flannery. *Numerical Recipes 3rd Edition: The Art of Scientific Computing*. Cambridge University Press, New York, 3rd ed., 2007.

[98] C. E. Rasmussen and C. K. I. Williams. *Gaussian Processes for Machine Learning*. MIT Press, Cambridge, 2006.

[99] L.A. Rastrigin. The convergence of the random search method in the extremal control of a many parameter system. *Automation and Remote Control*, 24(10):1337–1342, 1963.

[100] H.H. Rosenbrock. An automatic method for finding the greatest or least value of a function. *The Computer Journal*, 3(3):175–184, 1960.

[101] R. D. Ruth. A canonical integration technique. *IEEE Transactions on Nuclear Science*, 30(4):2669–2671, Aug 1983.

[102] J. Safranek. Experimental determination of storage ring optics using orbit response measurements. *Nucl. Instrum. Methods Phys. Res, Section A*, 388(1):27 – 36, 1997.

[103] J. Safranek, C. Limborg, A. Terebilo, K. I. Blomqvist, P. Elleaume, and Y. Nosochkov. Nonlinear dynamics in a SPEAR wiggler. *Phys. Rev. ST Accel. Beams*, 5, 01 2002.

[104] J. Safranek, G. Portmann, and A. Terebilo. Matlab-based LOCO. In *Proceedings of EPAC'02*, pages 1184–1186, 2002.

[105] D. Sagan and D. Rubin. Linear analysis of coupled lattices. *Phys. Rev. ST Accel. Beams*, 2:074001, Jul 1999.

[106] M. Sands. The physics of electron storage rings: an introduction. SLAC-R-121, 1970.

[107] A. Scheinker, X. Huang, and J. Wu. Minimization of betatron oscillations of electron beam injected into a time-varying lattice via extremum seeking. *IEEE Transactions on Control Systems Technology*, 26(1):336–343, Jan 2018.

[108] A. Scheinker and M. Krstic. Minimum-seeking for CLFs: Universal semiglobally stabilizing feedback under unknown control directions. *IEEE Transactions on Automatic Control*, 58(5):1107–1122, May 2013.

[109] X. Shen, S. Y. Lee, M. Bai, S. White, G. Robert-Demolaize, Y. Luo, A. Marusic, and R. Tomas. Application of independent component analysis to ac dipole based optics measurement and correction at the Relativistic Heavy Ion Collider. *Phys. Rev. ST Accel. Beams*, 16:111001, Nov 2013.

[110] V. Smaluk, X. Yang, W. Guo, Y. Hidaka, G. Wang, Y. Li, and L. Yang. Experimental crosscheck of algorithms for magnet lattice correction. In *Proceedings of IPAC'16*, pages 3400–3402, 2016.

[111] C. Steier, D. Robin, L. Nadolski, W. Decking, Y. Wu, and J. Laskar. Measuring and optimizing the momentum aperture in a particle accelerator. *Phys. Rev. E*, 65:056506, May 2002.

[112] Till Straumann. Employing conformal mapping techniques for BPM geometry analysis. unpublished, 2004.

[113] Y. Tan, W.H. Moase, C. Manzie, D. Nesic, and I.M.Y. Mareels. Extremum seeking from 1922 to 2010. In *Proceedings of the 29th Chinese Control Conference*, pages 14–25, 2010.

[114] A. Terebilo. Accelerator modeling with Matlab Accelerator Toolbox. In *Proceedings of PAC2001*, pages 3203–3205, Chicago, IL, 2001.

[115] K. Tian, J. Safranek, and Y. Yan. Machine based optimization using genetic algorithms in a storage ring. *Phys. Rev. ST Accel. Beams*, 17:020703, Feb 2014.

[116] R. Tomas, O. Brüning, M. Giovannozzi, P. Hagen, M. Lamont, F. Schmidt, G. Vanbavinckhove, M. Aiba, R. Calaga, and R. Miyamoto. CERN Large Hadron Collider optics model, measurements, and corrections. *Phys. Rev. ST Accel. Beams*, 13:121004, Dec 2010.

[117] Rogelio Tomas, Masamitsu Aiba, Andrea Franchi, and Ubaldo Iriso. Review of linear optics measurement and correction for charged particle accelerators. *Phys. Rev. Accel. Beams*, 20:054801, May 2017.

[118] Chun-xi Wang. Untangling mixed modes in model-independent analysis of beam dynamics in circular accelerators. *Phys. Rev. ST Accel. Beams*, 7:114001, Nov 2004.

[119] Chun-xi Wang, Vadim Sajaev, and Chih-Yuan Yao. Phase advance and β function measurements using model-independent analysis. *Phys. Rev. ST Accel. Beams*, 6:104001, Oct 2003.

[120] Andrzej Wolski. Alternative approach to general coupled linear optics. *Phys. Rev. ST Accel. Beams*, 9:024001, Feb 2006.

[121] X. Huang. Lecture Notes for Accelerator Physics, USPAS, June 2017.

[122] Lingyun Yang, Yongjun Li, Weiming Guo, and Samuel Krinsky. Multiobjective optimization of dynamic aperture. *Phys. Rev. ST Accel. Beams*, 14:054001, May 2011.

[123] X. Yang, G. Ganetis, Y. Hidaka, T.V. Shaftan, V.V. Smaluk, G.M. Wang, L.-H. Yu, and P. Zuhoski. Online optimization of NSLS-II dynamic aperture and injection transient. In *Proceedings of IPAC'19*, Melbourne, Australia, 2019.

[124] X. Yang, V. Smaluk, L.H. Yu, Y. Tian, and K. Ha. Multi-frequency AC LOCO: a fast and precise technique for lattice correction. In *Proceedings of IPAC'17*, pages 831–834, 2017.

[125] Xi Yang and Xiaobiao Huang. A method for simultaneous linear optics and coupling correction for storage rings with turn-by-turn beam position monitor data. *Nucl. Instrum. Methods Phys. Res, Section A*, 828:97–104, 2016.

[126] L. H. Yu, R. Biscardi, J. Bittner, A. M. Fauchet, S. Krinsky, R. J. Nawrocky, J. Rothman, 0. V. Singh, and K. M. Yang. Real time global orbit feedback system for NSLS X-ray ring. In *Proceedings of PAC'91*, pages 2542–2545, 1991.

[127] L. H. Yu, E. Bozoki, J. Galayda, S. Krinsky, and G. Vignola. Real time harmonic closed orbit correction. *Nucl. Instrum. Methods Phys. Res, Section A*, 284(2):268–285, 1989.

[128] Tong Zhang, Xiaobiao Huang, and Tim Maxwell. Linear optics correction for linacs and free electron lasers. *Phys. Rev. Accel. Beams*, 21:092801, Sep 2018.

[129] A. G. Zhilinskas. Single-step Bayesian search method for an extremum of functions of a single variable. *Cybernetics and Systems Analysis*, 11:160–166, 01 1975.

Index

Printed in the United States
by Baker & Taylor Publisher Services

Printed in the United States
by Baker & Taylor Publisher Services